외국인을 위한 기초 PLC

(XG5000 / GX_Works2)

최은혁 지음

Basic PLC for International Learners

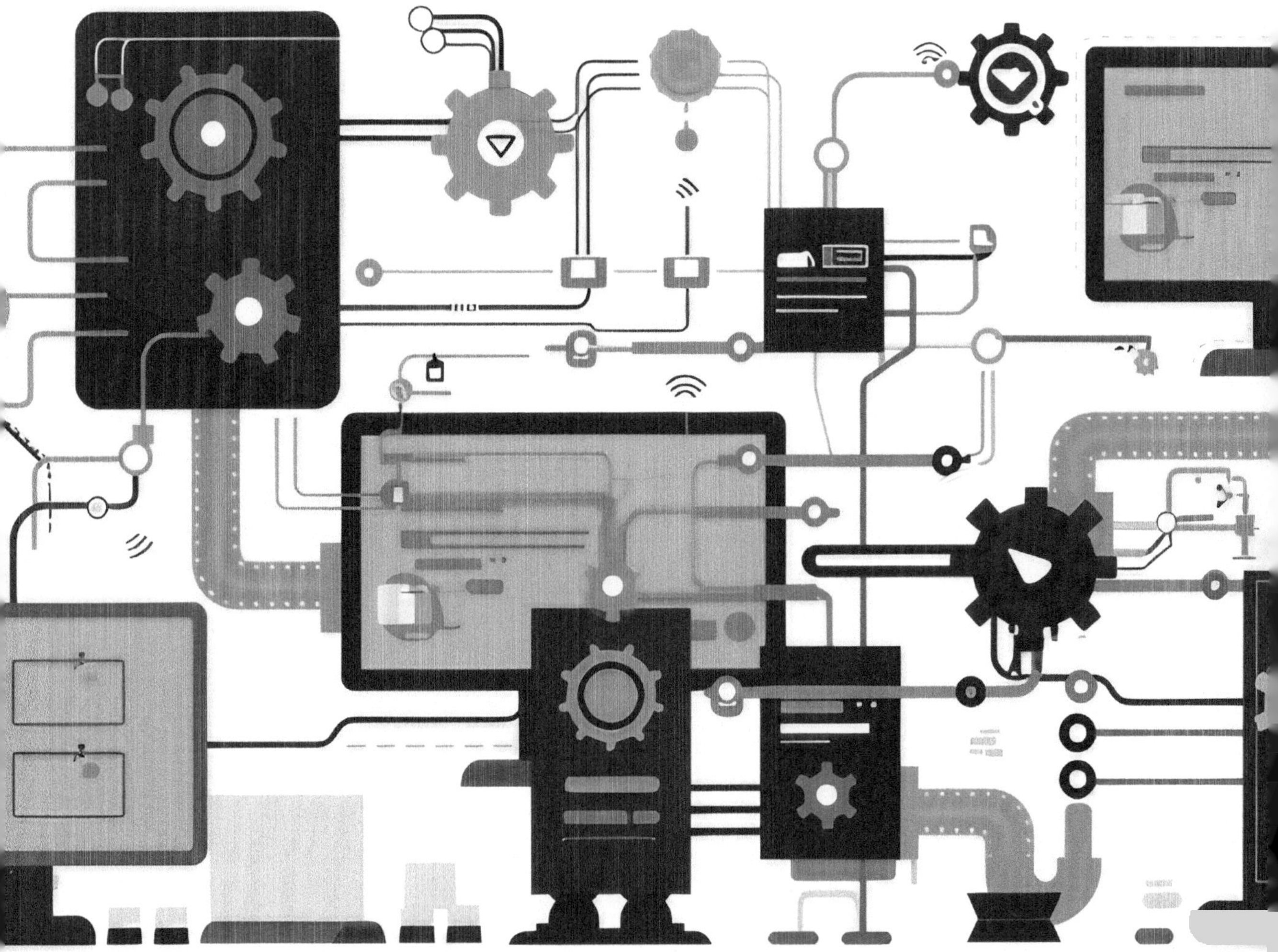

도서출판 홍릉 | 50Th

저자소개

최은혁

현재 경일대학교 전기공학과 교수로 영남대학교 공학박사 학위를 취득하였다. 주요 연구분야는 고전압, 전자파공학 및 스마트팩토리이다. 저서로는「NCS를 기반으로한 일학습병행제 관련 전기·전기설계」(한국인력공단, 2015),「제어시스템 실무 및 PLC제어」(한국인력공단, 2015),「처음만나는 전기기기」(한빛아카데미, 2016),「처음만나는 전력공학」(한빛아카데미, 2016),「스마트팩토리 PLC(GM-Win)」(홍릉과학출판사, 2018),「전기기사·산업기사를 위한 공업수학」(도서출판 홍릉, 2021),「처음만나는 전기기기-2판」(한빛아카데미, 2021) 등이 있다.

외국인을 위한
기초 PLC
(XG5000 / GX_Works2)

인 쇄: 2025년 12월 22일 초판 1쇄
발 행: 2025년 12월 29일 초판 1쇄

저 자: 최은혁
발행인: 송 준
발행처: 도서출판 홍릉
주 소: 01093 서울시 강북구 인수봉로 50길 10
등 록: 1976년 10월 21일 제5-66호

전 화: 02-999-2274~5, 903-7037
팩 스: 02-905-6729
e-mail: hongpub@hongpub.co.kr
http://www.hongpub.co.kr
ISBN: 979-11-7570-017-8(93560)

정 가: 19,000원

차 례

PART I

XGT PLC(CPU XGI, XG5000

PART
II

XMITSUBUBISHI PLC(GX-Works2)

PART I

XGT PLC (CPU XGI, XG5000)

1.
XGT PLC
(CPU XGI)

What is a PLC?

- PLC (Programmable Logic Controller) is a highly autonomous control device that replaces the functions of relays, timers, and counters previously used in control panels with semiconductor components such as ICs and transistors. It adds arithmetic functions to basic sequence control, enabling program-based control.
- According to the National Electrical Manufacturers Association (NEMA) in the United States, a PLC is defined as "an electronic device that uses programmable memory to perform specific functions such as logic, sequencing, timing, counting, and arithmetic through digital or analog input/output modules, and controls various types of machinery or processes through digital operations."

PLC Functions

- Sequencing: Digital input/output control
- Logic operations: AND, OR, NOT, XOR, etc.
- Arithmetic operations: Various arithmetic operations, PID operations
- Timer: Time delay, accumulated time, time-specified pulse
- Counter: Counting the number of input pulses
- Analog input: Current, voltage, temperature
- Analog output: Current, voltage
- High-speed pulse input: Counting the number of high-speed input pulses
- Pulse train output: Stepping, servo motor control
- Communication: Transfer of data using communication protocols

Birth of the PLC

In 1967, General Motors (GM) in the United States presented the requirements for a new control device to simplify the wiring replacement process when changing production models on the automobile manufacturing line. These conditions led to the development of the PLC.

- Programmable: Capable of programming and modifying programs
- Operates in harsh industrial environments
- Input: Capable of receiving AC 120V signals
- Output: Able to directly drive actuators
- Cost-effective
- Expandable
- Compact size
- Reliable and easy to maintain
- Communication capabilities
- Programmable memory of more than 4K steps

Comparison of Relay (Analog) Control and PLC

Category	Relay Control	PLC Control
Sequence Control	○	○
Timer	○	○
Counter	○	○
Logic Operations	×	○
Arithmetic Operations	×	○
Analog	×	○
High-speed Pulse	×	○
Communication	×	○

Applications of PLC

With the increasing demand for automation and high efficiency in equipment, the application scope of PLCs is expanding. In particular, the demand for PLCs in factory automation and Flexible Manufacturing Systems (FMS) has evolved from replacing medium-to-large relay control panels in the past to currently being used in high-function, high-speed systems. PLCs are now applied in a wide range of areas, from small machine tools to large-scale system installations.

Applications	Control Targets
Steel and Iron Industry	Control of loading operations, raw material transportation, rolling lines
Chemical and Textile Industry	Control of raw material shipping, raw material input, reaction temperature, dyeing lines
Automotive Industry	Control of transfer lines, assembly lines, painting lines
Machinery Industry	Control of machine tools, industrial robots, servo motors
Logistics Industry	Control of automated warehouses, conveyors, loading equipment, transport lines
Water and Sewage	Control of water purification plants, sewage treatment plants, water supply and drainage
Factory Equipment	Power monitoring, monitoring, and control of various utilities

PLC Structure

1. Hardware Structure

(1) Overall Configuration

- Central Processing Unit (CPU): Includes a microprocessor and memory
- Input/Output Section: Connects signals with external devices
- Power Supply Section: Supplies power to each component
- Peripheral Devices: Used to write programs into the PLC's memory

2. PLC Input/Output Section

The input/output section of a PLC is used by directly connecting it to external devices in the field. While the internal components of the PLC use a DC +5V power supply (TTL level), the input/output section operates at different voltage levels. Therefore, the connection (interface) between the internal PLC and the input/output section is a critical factor for system stability.

Category		PLC Control
Input Section	Operation Input	Various switches
	Detection Input	Various sensors
Output Section	Alarm Output	Lamp, buzzer
	Drive Output	Valve, brake, switchgear

3. Circuit Configuration

When controlling the electrical circuit shown in the left diagram using a PLC, the switches are categorized as inputs and the lights as outputs, as shown in the right diagram, and the circuit must be configured separately. The input and output are configured separately, and the connections between them are made through the PLC's program. Since the PLC program is handled through software, it can be easily edited and modified. Therefore, once the input and output are wired, all logic circuits can be freely changed and modified through the program.

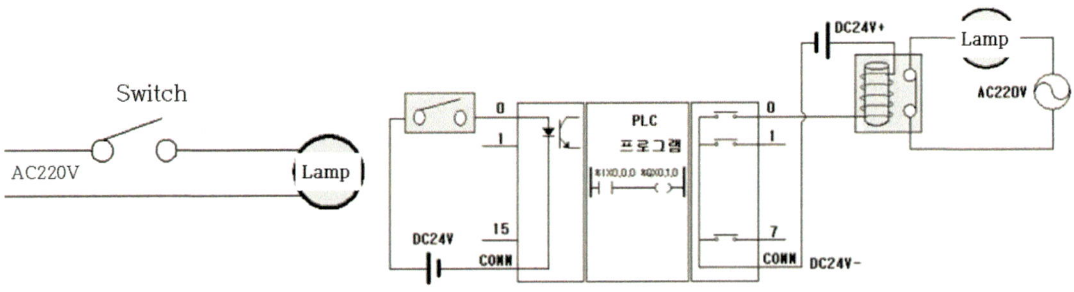

Arithmetic Processing in Software

1. Serial Processing and Parallel Processing

The fundamental difference between PLC sequencing and relay sequencing lies in the "serial processing and parallel processing" as shown in the diagram. PLC operates using a serial processing method, where it sequentially processes the program stored in memory. In contrast, relay sequencing operates in parallel, with multiple circuits functioning simultaneously through electrical signals. Therefore, when observing a PLC at any given moment, it is only performing one task at a time.

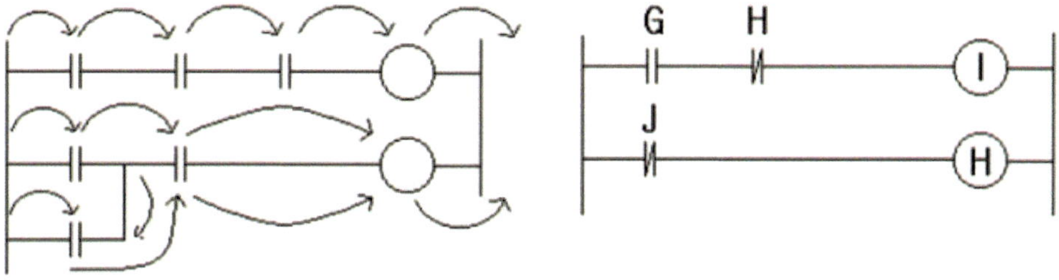

〈 Serial Processing 〉

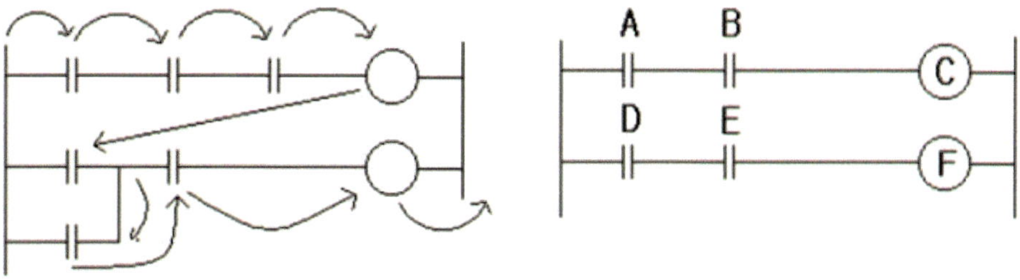

〈 Parallel Processing 〉

2. Limitation on the Number of Contacts Used

In relay systems, there is generally a limit to the number of contacts each relay can have. Therefore, when creating a relay sequence, it is necessary to conserve the number of contacts as much as possible. In contrast, a PLC does not have a limit on the number of times the same contact can be used. This is because the ON/OFF status of the same contact is stored in a designated memory, and during processing, the PLC reads the information from memory and executes the operation.

PLC	Relay
- Unlimited usage of the same contact within the program's capacity - The ON/OFF status of the same contact is stored in designated memory - During processing, the PLC reads the information from memory and executes the operation	- Limit on the number of contacts that each can have - Restriction on the number of times the same contact can be used

3. Contact and coil position limitations

In PLC sequencing, there are certain rules that do not exist in relay sequencing. One of these is the prohibition of placing contacts after a coil. In PLC sequences, coils must always be placed on the right side of the bus. Additionally, in PLC sequences, the signal must always flow from left to right.

Therefore, unlike relay sequences, PLC sequences prohibit circuits that flow from right to left or circuits that flow vertically.

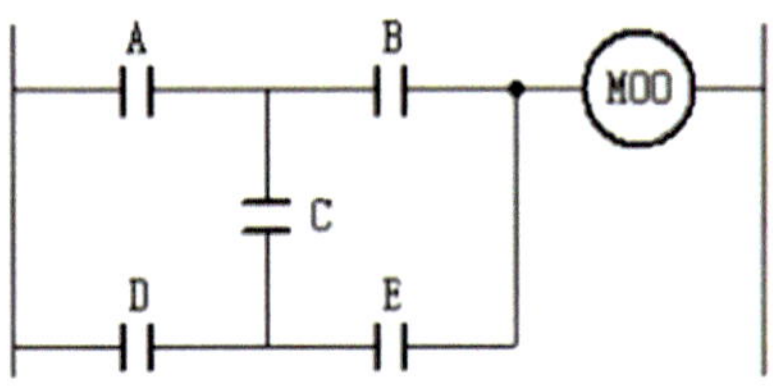

Vertical flow is prohibited

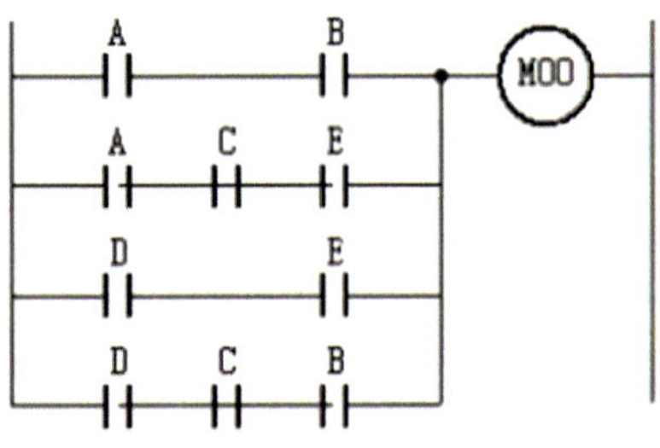

Always from left to right

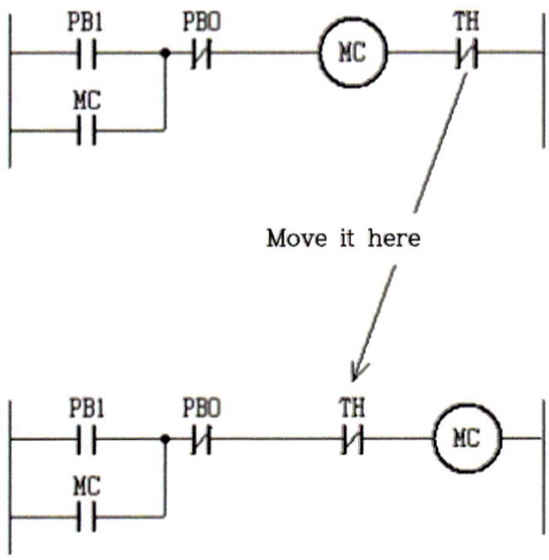

As shown in the diagram below, the PLC executes instructions starting from the top-left command, moving to the right, and then proceeding downward one instruction at a time. After executing all the instructions in this manner, it will finally encounter the END command.

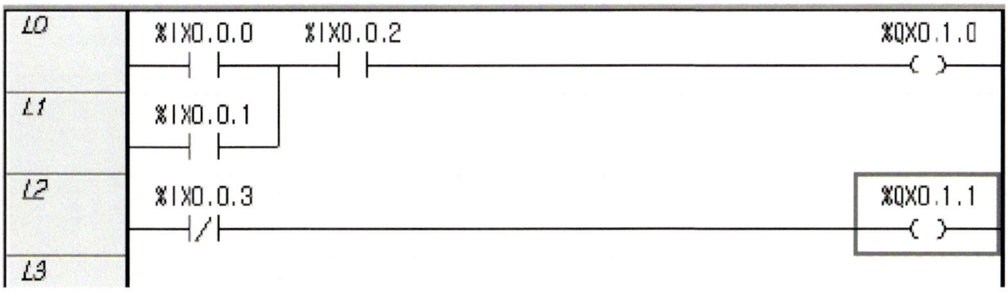

프로그램 종료(END)

Operating Principle of PLC

A PLC controls external input and output devices connected to its main unit according to the user's program. Therefore, for accurate operation, it is essential to understand the correct wiring of the input/output devices, the program, and the characteristics of PLC control.

1. PLC Programming Languages

The currently used programming languages include IL (Instruction List), ST (Structured Text), Ladder, FBD (Function Block Diagram), and SFC (Sequential Function Chart). XGI PLC supports four languages: IL (Instruction List), ST (Structured Text), Ladder, and SFC (Sequential Function Chart).

(1) IL (Instruction List)

In MASTER-K PLC, the IL is referred to as mnemonic instructions and is a text-based language in the form of assembly language. In MASTER-K, it was mainly used for simple on-site logic programming with a handy loader. However, with the advent of laptops replacing this function, the necessity of programming in IL has diminished, making it a rarely used language today.

No.	Instruction	Input Parameter	Variable	Variable Description

번호	명령어	입력 파라미터	변수	변수 설명문
1	LOAD		%IX0.0.0	
2	OR		%IX0.0.1	
3	AND		%IX0.0.2	
4	OUT		%QX0.1.0	
5	LOAD NOT		%IX0.0.3	
6	OUT		%QX0.1.1	

(2) Ladder (Ladder Diagram):

Ladder is a graphical, diagram-based language similar to relay logic, where logic is expressed in the form of a ladder without explicitly showing the power connections. It is currently the most widely used PLC programming language.

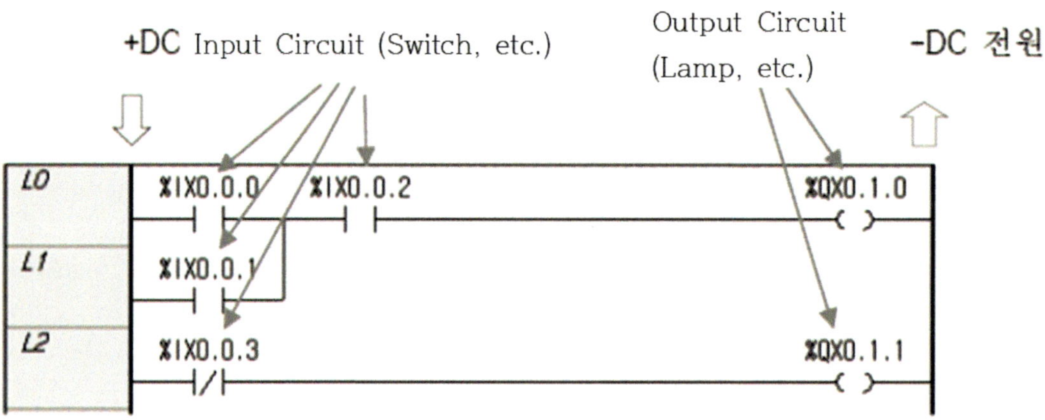

2. Understanding PLC Operation

(1) Basic PLC Symbols (Instructions)

PLC includes input switches and output coils that are similar to relay logic. These symbols represent the basic building blocks of PLC logic:

Category	Relay Logic	PLC Logic	Description
A Contact	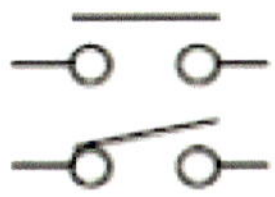		Normally Open (NO) Contact N.O(Normally Open) PLC : Enter the ON/OFF status of external input and internal output.
B Contact	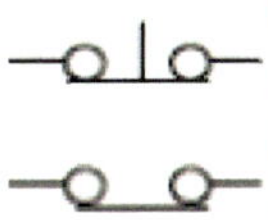		Normally Closed (NC) Contact N.C (Normally Closed) PLC : Input the ON/OFF status of external inputs and internal outputs in the reversed state
C Contact	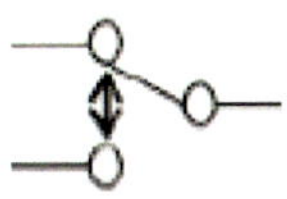	None	In a PLC, mixed-type a and b contacts are expressed through a combination of logic
Output Coil		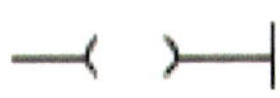	Output the contact based on the previous operation result.
Application Command	None	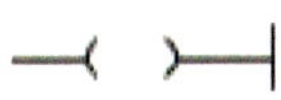	Execute the PLC application command

Overview of XGI

1. IEC Standard Languages

The IEC standardized languages for PLCs consist of two graphic-based languages, two text-based languages, and SFC (Sequential Function Chart).

(1) Graphic Languages

ⓐ LD (Ladder Diagram): A language that expresses relay logic.

ⓑ FBD (Function Block Diagram): A language that represents programs by connecting functional blocks.

(2) Text-Based Languages

ⓐ IL (Instruction List): A language in the form of assembly language.

ⓑ ST (Structured Text): A high-level language similar to Pascal.

(3) SFC (Sequential Function Chart): A program development method that unfolds sequentially, similar to a flowchart.

Currently, XGI PLC supports IL, LD, and SFC languages.

2. Common Elements

The program components of XGI/XGR/XEC PLCs (programs, functions, function blocks) can be written in different languages such as LD, SFC, and ST. However, these languages share common syntax elements that are used universally across them.

(1) Expression

1) Identifiers

- Any combination of letters, digits, and underscores (_) that starts with a letter or an underscore can be used as an identifier.
- Identifiers are used as variable names.
- Identifiers must not contain spaces.
- When used as variable or instance names, identifiers are typically not restricted to Korean, English, or Chinese characters.
- In the case of English letters, uppercase and lowercase are not distinguished, and all are recognized as uppercase.

Type	Usage
Uppercase Letters and Numbers	IW210, IW215Z, QX75, IDENT
Underscore Character	LIM_SW_2, LIMSW5, ABCD, AB_CD
Uppercase Letters and Numbers Starting with an Underscore	_MAIN, _12V7, _ABCD

Representation of Variables

The data used in a program holds values, and during program execution, there are constants whose values do not change and variables whose values do change. In order to use variables in program components such as program blocks, functions, and function blocks, the representation of variables must first be explained.

Variable Representation Methods

▷ Direct Variables: No need to declare variables (traditional PLC method)

▷ Named Variables: Variable declaration is required

The first method of variable representation is the direct variable method, where the user does not assign a name and uses identifiers of memory areas already designated by the manufacturer. The second method is the named (symbolic) variable method, where the user assigns and uses a specific name for the variable.

1. Direct Variables

(1) Identifiers are represented without separate variable declarations.

(2) Identifiers and addresses for memory areas designated by the manufacturer are used.

(3) The user represents the identifier location without declaring variable names or types.

(4) The size of input/output variables and internal memory variables varies depending on the model.

(5) Starts with a percent symbol (%).

(6) Follows with a location prefix (Input: %I, Output: %Q, Internal: %M).

(7) Uses a size prefix (Bit: X, Byte: B, Word: W, Double Word: D, Long Word: L)

<table>
<tr><th>Type</th><th>Usage</th><th>Type</th><th>Usage</th></tr>
<tr><td>Input</td><td>%IX0.0.0, %IX0.0.1</td><td rowspan="2">Internal</td><td>%MX100, %MW100</td></tr>
<tr><td>Output</td><td>%QX0.1.0, %QW0.1.0</td><td>%MB50.3, %MD50.31</td></tr>
</table>

2. Named Variable

(1) The variable name, type (attribute), data type, and memory allocation are declared and used.

(2) Variable names can be up to 8 characters for Korean/Chinese characters and up to 16 characters for English.

(3) A combination of Korean, English, numbers, and the underscore (_) can be used.

(4) English letters are recognized as uppercase.

(5) Variable names must not contain spaces.

Type	Usage
Korean, Numbers	Motor1, Push_Detection, Digital Switch1
English, Numbers	BCD, AUTO_SOL_VAL, MOTOR_2

3. Declaration Procedure for Named Variables

The procedure for declaring named variables is as follows:

(Variable type setting ⇒ Data type specification ⇒ Memory allocation)

(1) Setting the Variable Type (Attribute)

The attribute is set according to the purpose of the variable as follows:

① VAR: A general variable that can be read and written.

② VAR-RETAIN: A variable whose value is retained after power failure and recovery.

③ VAR-CONSTANT: A read-only variable.

④ VAR-EXTERNAL: A variable used as an external (GLOBAL) variable.

A global variable is a variable that is commonly used across multiple programs within a single project.

4. Specifying the Data Type

The data type represents the inherent properties of the data. Data types can be broadly categorized into numeric (ANY_NUM) and bit status (ANY_BIT). A typical example of numeric data is an integer

(INT), which can be counted and used in arithmetic operations. Examples of integers include the current value of a counter or the value from an A/D (analog input) conversion.

Bit status types include BOOL (1 bit), BYTE (a series of 8 bits), and WORD (a series of 16 bits). These represent the ON/OFF status of bit series and can be used in logical operations. Examples of bit status include the ON/OFF status of an input switch or the on/off state of an output lamp. Bit status types cannot be used in arithmetic operations, but they can be converted into numerical values using a type conversion function, allowing arithmetic operations. BCD represents decimal numbers in 4-bit binary code and is considered a bit status type (ANY_BIT).

5. Memory Allocation for Named Variables

Memory allocation for named variables can be either automatic or user-defined.

(1) Automatic Allocation

① The compiler automatically assigns variable locations within the internal memory area.

② It temporarily stores signal states or relays signals during internal operations.

③ It is used for instance designation of function blocks, such as timers or counters.

(2) User-Defined

① The user directly specifies the location by using variables (%I, %Q, %M).

② Input Variables: Use %I variables.

③ Output Variables: Use %Q variables.

④ Internal Memory Variables: Use %M variables.

⑤ The format for expressing user-defined memory allocation is the same as the method used for specifying direct variables.

Address Allocation

1. Input/Output Memory Allocation

(1) Format of Input/Output Memory Allocation

%I X . N1 . N2 . N3
① ② ③ ④ ⑤ (N1, N2, N3 is a number)

① Location Prefix - Type of Variable

► I: Input area

► Q: Output area

② Size Prefix - Size of Memory Space Occupied by the Variable

► X: Size of 1 bit (the "X" character can be omitted).

► B: Size of 1 byte (8 bits).

► W: Size of 1 word (16 bits).

► D: Size of 1 double word (32 bits).

► L: Size of 1 long word (64 bits).

③ Base Number

The base where the CPU is mounted is called Base 0. When an expansion system is configured, the base number increases in sequence according to the connection order to the main base.

④ Slot Number

For the main base, the slot number starts at 0 for the right side of the CPU, increasing by 1 as it moves to the right. For expansion bases, the slot number starts at 0 for the right side of the power supply and increases by 1 as it moves to the right.

⑤ Size Prefix Number

This indicates the sequence number of the size prefix unit when the contacts mounted in the slot are divided into size prefix units starting from bit 0.

2. Internal Memory Address Allocation

The allocation of internal memory follows the same basic method as the input/output memory allocation described above, but without specifying base numbers and slot numbers. There are two ways to represent internal memory:

(1) Representation by Size Prefix Unit

<u>%M</u> <u>B</u> <u>N1</u> (N1는 숫자)
① ② ③ (N1 is a number)

① %M in item 1 is the location prefix that represents internal memory.

② Item 2 is the size prefix, which is the same as in input/output memory:

- ► X: Size of 1 bit (the "X" character can be omitted).
- ► B: Size of 1 byte (8 bits).
- ► W: Size of 1 word (16 bits).
- ► D: Size of 1 double word (32 bits).
- ► L: Size of 1 long word (64 bits).

③ Item 3 indicates the size prefix number.

(2) Representation of Bits Using the Size Prefix

<u>%M</u> <u>B</u> <u>N1</u> . <u>N2</u> (N1, N2는 숫자)
① ② ③ ④ (N1, N2 is a number)

① %M in item 1 is the location prefix representing internal memory.

② Item 2 is the size prefix, where B, W, D, and L can be used, excluding X:

- ► B: Size of 1 byte (8 bits).
- ► W: Size of 1 word (16 bits).
- ► D: Size of 1 double word (32 bits).
- ► L: Size of 1 long word (64 bits).

③ Item 3 indicates the size prefix number.

④ Item 4 represents the bit number.

XG5000 Program

1. Screen Layout

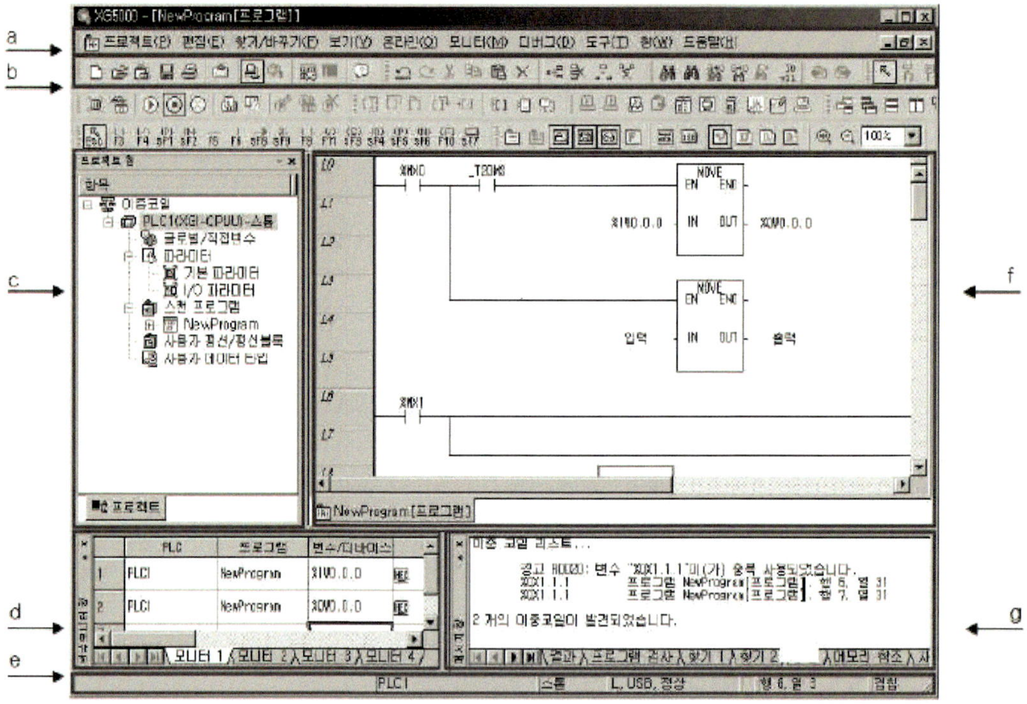

[Dialog Box Description]

ⓐ Menu: The basic menu for the program.

ⓑ Toolbar: Allows easy access to menu functions.

ⓒ Project Window: Displays the components of the currently open project.

ⓓ Variable Monitor Window: Allows monitoring of registered variables.

ⓔ Status Bar: Displays the status of XG5000, connected PLC information, etc.

ⓕ Edit Window: The current LD (Ladder Diagram) edit window is visible.

ⓖ Message Window: Displays various messages that occur while using XG5000.

2. Menu Structure

When a menu is selected, commands will appear. You can execute a command by selecting it with the mouse or keyboard. For menus with shortcut keys (e.g., Ctrl+X, Ctrl+C), you can use the shortcuts to directly select the command.

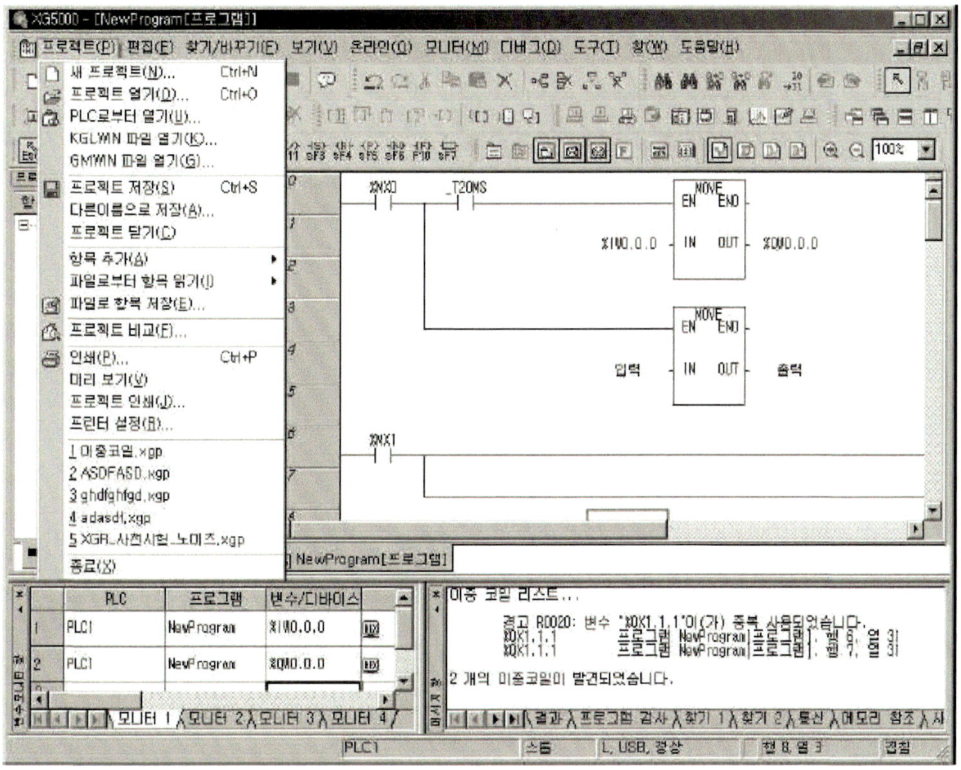

3. Toolbar

XG5000 provides frequently used menus in the form of shortcut icons. You can execute a desired tool by clicking it with the mouse.

4. Status Bar (At the bottom of the screen)

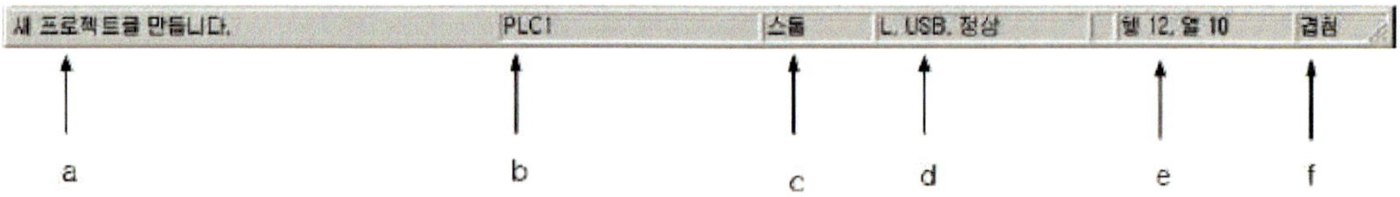

ⓐ Command Description: Displays a description of the selected menu, command, or toolbar item where the mouse is positioned.

ⓑ PLC Name: Displays the name of the selected PLC. If a project contains multiple PLCs, online-related commands will apply to the PLC shown here.

ⓒ PLC Mode Display: Shows the mode of the PLC. If a project contains multiple PLCs, the mode of the selected PLC will be displayed.

ⓓ Warning Display: Indicates the abnormal status (error) of the PLC.

ⓔ Cursor Position Display: Shows the position of the cursor when editing a program.

ⓕ Mode Display: Displays either insert mode or overwrite mode.

5. Opening and Closing a Project

(1) Opening a Project

① Select the menu option [Project] - [Open Project].

② Choose the profile and click the Open button.

③ The file extension for XG5000 project files is "xgp." After selecting the project file in the open dialog box, you can check the project description in the description area.

(2) Closing a Project

① Select the menu option [Project] - [Close Project]. There is no default shortcut key set for this action.

② If the project has been edited but not saved, the following message will appear.

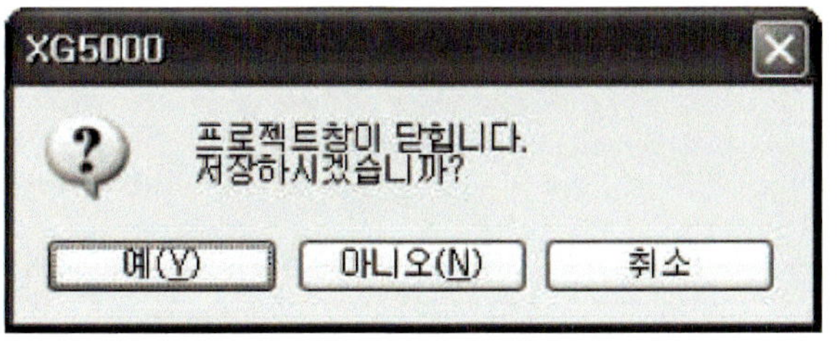

③ If you wish to save, click the Yes button.

6. Saving a Project

① Select the menu option [Project] - [Save Project].

② If an asterisk (*) appears next to the project name in the project window, it indicates that the current project has been edited.

7. Configuring Options - Dialog Box

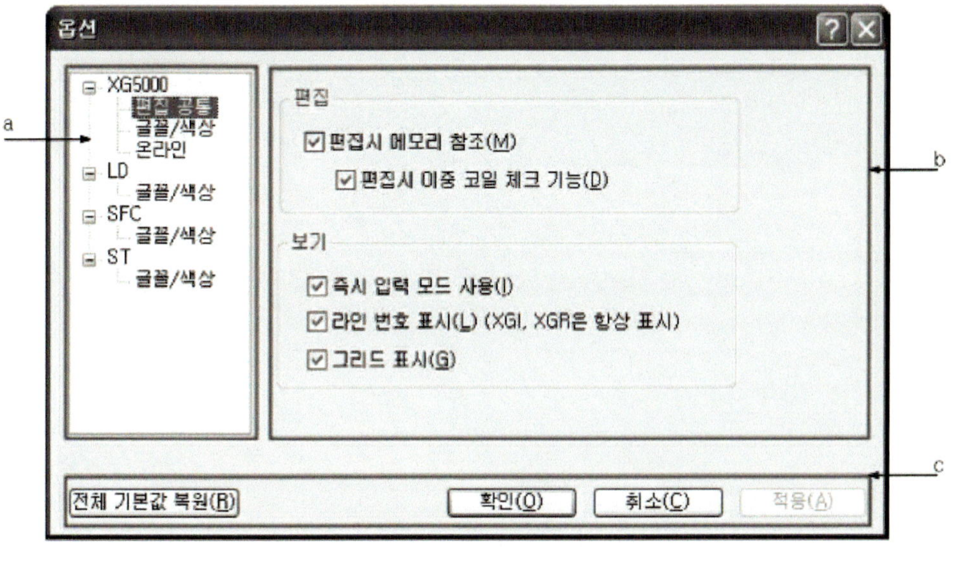

ⓐ Category: A tree-structured classification of XG5000 options that apply to the entire program, as well as options that can be applied to specific languages.

ⓑ Settings Content: When a category from (a) is selected, the relevant settings for that category are displayed.

ⓒ General Buttons: Common buttons that apply to all categories, regardless of the selected category.

ⓓ Restore All Defaults Button: Used to restore all options to their default se

8. Configuring Options - Project

(1) Procedure

① Select the menu option [Tools] - [Options].

② In the options dialog box, select XG5000.

(2) Dialog Box

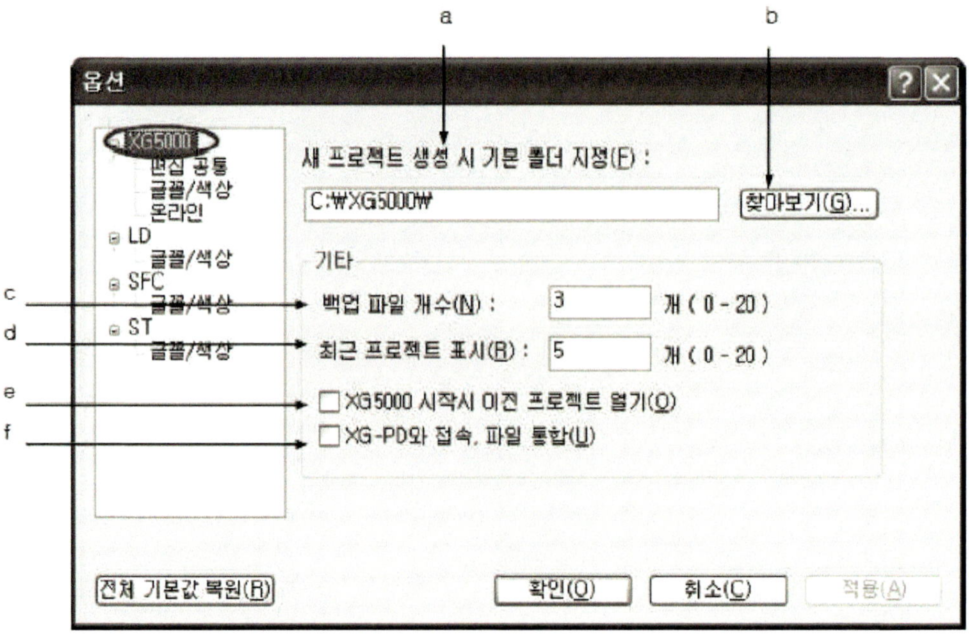

ⓐ Specify Default Folder for New Project Creation: The location where new projects will be created.

ⓑ Browse: Search for a folder.

ⓒ Set the Number of Backup Files for Project Recovery: You can set up to 20 backup files for recovering a project.

ⓓ Set the Number of Recent Projects Displayed: Configure how many recently opened projects will appear in the [Project] - [Recent Projects] menu (up to 20).

ⓔ Automatically Open the Most Recent Project: When checked, XG5000 will automatically open the most recent project when starting.

ⓕ Synchronize XG-PD with XG5000: When running XG-PD through the XG5000 menu, the connection options and PLC name from XG5000 will be displayed in XG-PD.

9. Projet

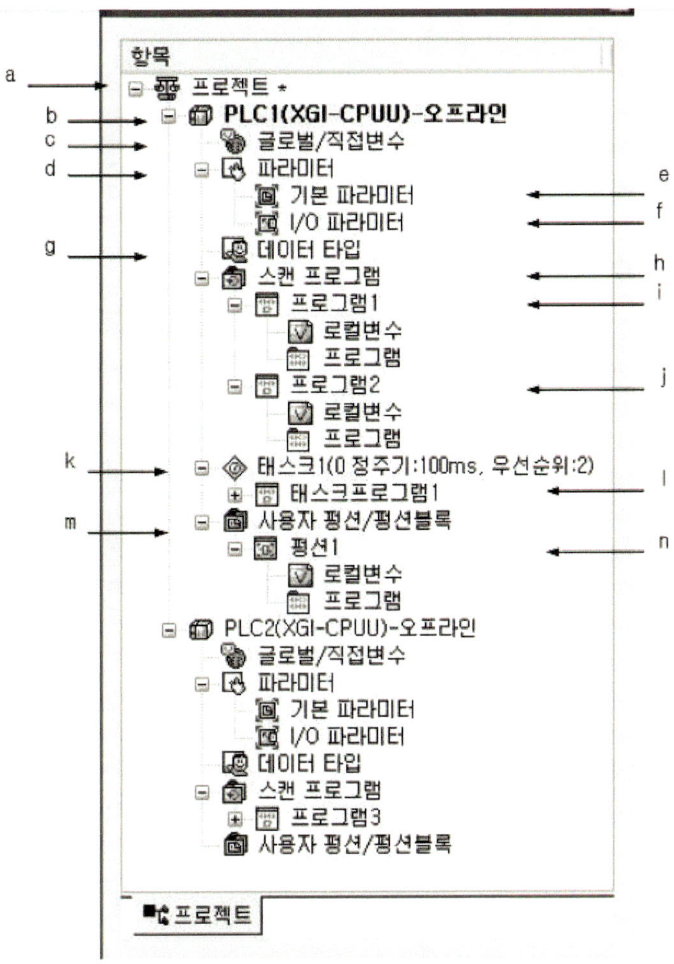

ⓐ Project: Defines the entire system. A project can include multiple related PLCs.

ⓑ PLC: Represents a system corresponding to one CPU module.

ⓒ Global/Direct Variables: Allows editing and viewing of global variable declarations and direct variable descriptions.

ⓓ Parameter: Defines the operations and configuration of the PLC system.

ⓔ Basic Parameter: Defines basic operations.

ⓕ I/O Parameter: Defines the configuration of input/output modules.

ⓖ Data Type: Defines structure types.

ⓗ Scan Program: Defines programs that are always executed in the sub-items.

ⓘ Program 1: A continuously running program defined by the user.

ⓙ Program 2: Another continuously running program defined by the user.

ⓚ Task 1: A user-defined cyclic task.

ⓛ Task Program 1: A program executed based on Task 1 conditions.

ⓜ User Function/Function Block: Allows users to create functions/function blocks in the sub-items.

ⓝ Function 1: A function created by the user.

A project can include multiple PLCs. Managing multiple PLCs within one project is convenient, and after running XG5000, you can simultaneously connect to and monitor multiple PLCs.

XG5000 Project File Management

1. Creating a New Project

① Select the menu option [Project] - [New Project].

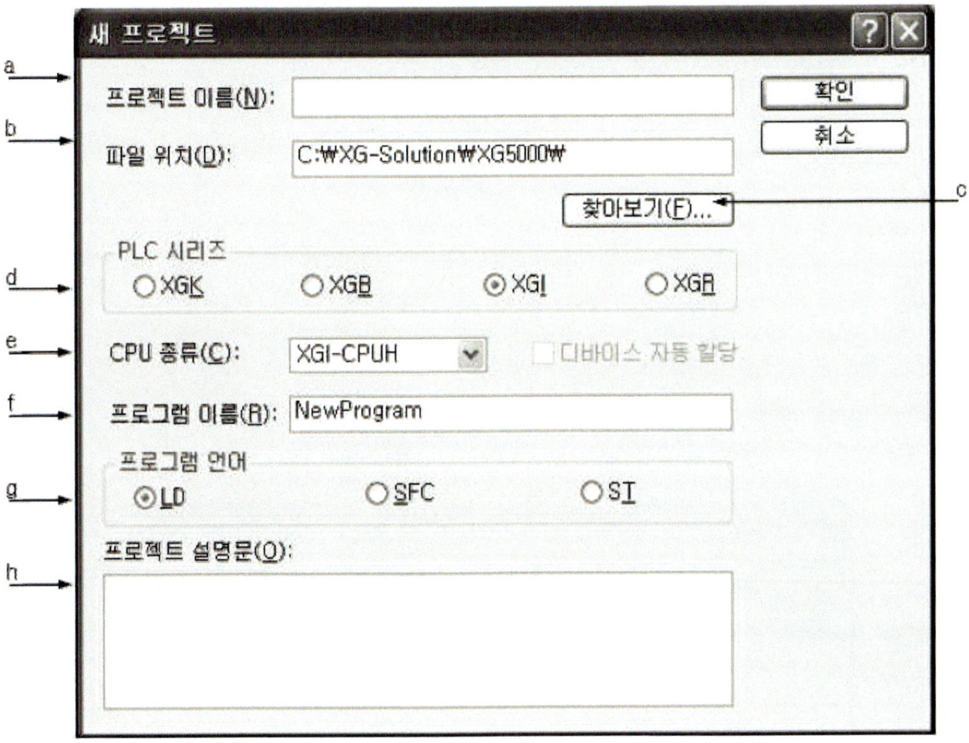

ⓐ Project Name: Enter the desired project name. This name becomes the project file name. The project file extension is "xgp."

ⓑ File Location: A folder will be created with the project name entered by the user, and the project file will be generated in that folder.

ⓒ Browse: Browse existing folders to specify the location of the project file.

ⓓ Project Type: Select between XGK format, XGB format, or XGI (IEC programming) format. Once selected, XGK, XGB, and XGI formats are not compatible with each other.

ⓔ PLC Type: Select the PLC model.

ⓕ Program Name: Enter the name of the default program to be included in the project.

ⓖ Programming Language: Select the language for the default program included in the project.

ⓗ Project Description: Enter a description for the project.

ⓘ Project File: When creating a new project, a folder with the same name as the project file will be created, and the project file will be generated inside it. The project file extension "xgp" is automatically added, even if not entered manually.

2. Opening a Project

① Select the menu option [Project] - [Open Project].

② When you select a project file, the description written by the user will appear in the description field.

③ This description can help with selecting the correct project.

④ After selecting the project file, click the Open button.

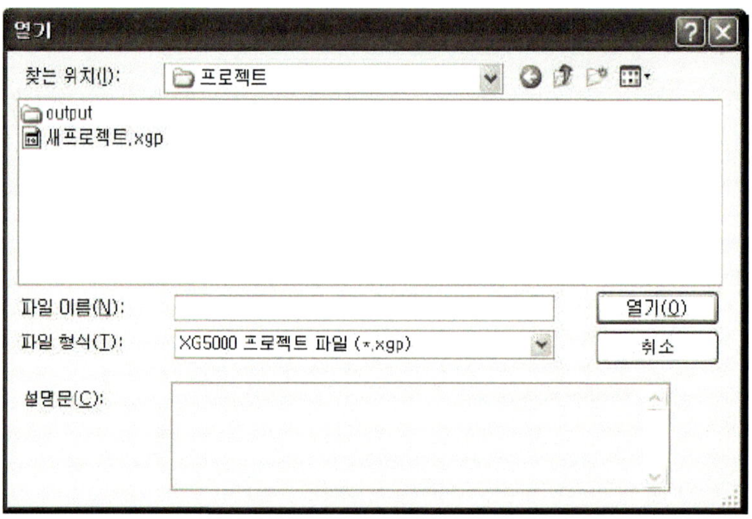

3. Saving a Project

① Select the menu option [Project] - [Save Project].

② If the project has been edited and needs to be saved, an asterisk ("*") will appear next to the project name in the project window.

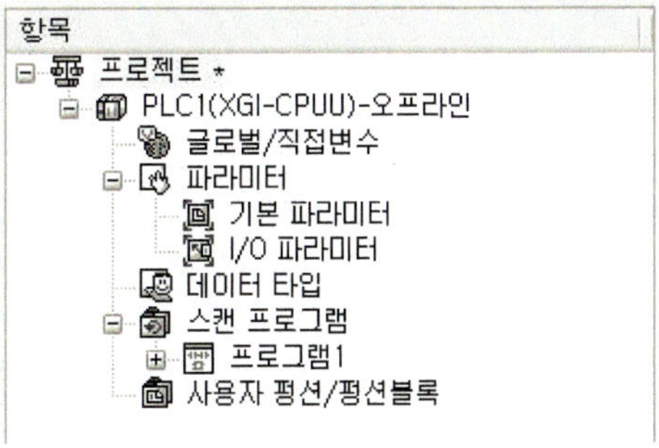

4. Save As

① Select the menu option [Project] - [Save As].

② Click the Confirm button.

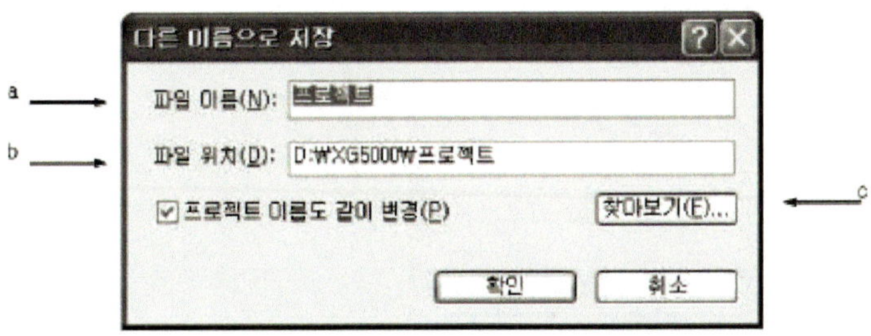

ⓐ Project Name: Enter the desired project name. This name becomes the project file name. The project file extension is "xgp."

ⓑ File Location: A folder with the same name as the project will be created automatically, and the project file will be generated in that folder.

ⓒ Browse: Browse existing folders to specify the location for the project file.

Variables / Description

1. Global / Direct Variables

Global/Direct variables consist of global variables, direct variable descriptions, and flags. Global variables allow you to declare variables that will be used in the program, or display the entire list of declared variables with a focus on variables.

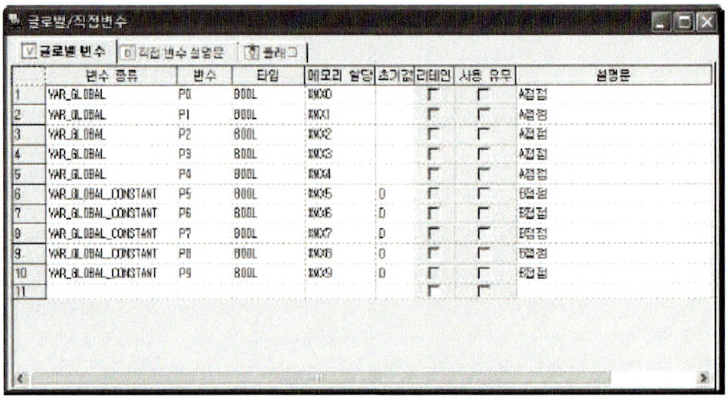

2. Direct Variable Description

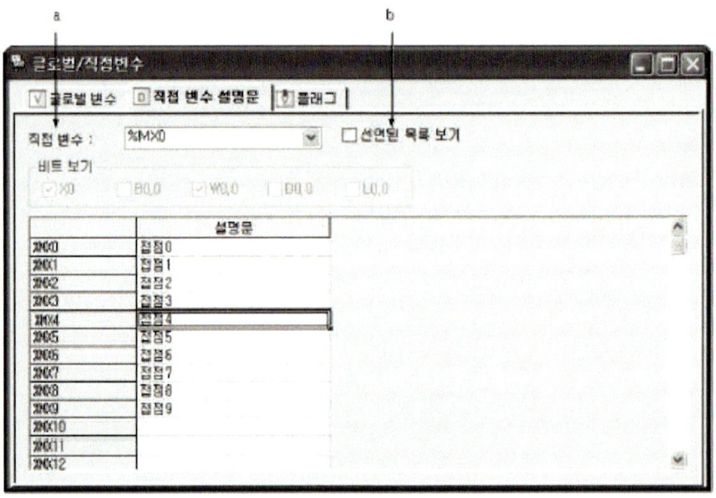

ⓐ Direct Variable: When the corresponding direct variable is entered, the data of the direct variable description is displayed.

ⓑ View Declared List: Displays the list that contains direct variable descriptions.

LD (Ladder Diagram) Editing

1. Restrictions

Item	Limitation	Limitation
Maximum number of points per line	The maximum number of points that can be input in one line	31 points
Maximum number of lines	The maximum number of editable lines	65,535 lines
Maximum number of lines for copy	The maximum number of lines that can be copied at one time	300 lines
Maximum number of lines for paste	The maximum number of lines that can be pasted at one time	300 lines

2. Editing Tools

To input LD editing elements, select the desired element from the LD toolbar and then click on the designated position with the mouse or use a shortcut key to begin.

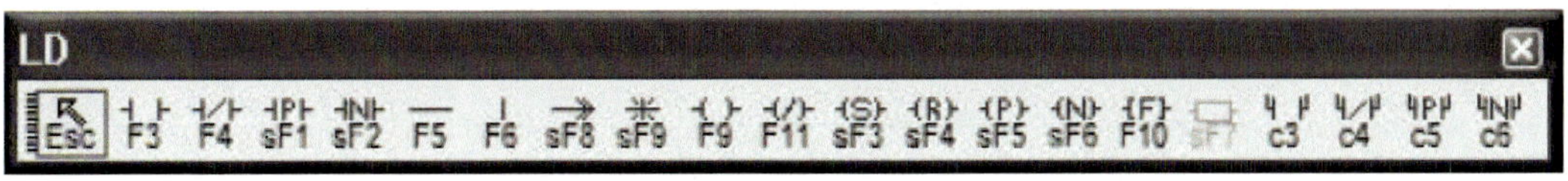

Symbol	Shortcut Key	Description
	Esc	Switch to selection mode
	F3	Normally open contact
	F4	Normally closed contact
	Shift + F1	Positive conversion detection contact
	Shift + F2	Negative conversion detection contact
	F5	Horizontal line

Symbol	Shortcut Key	Description
F6	F6	Vertical line
sF8	Shift + F5	Connection line
sF9	Shift + F6	Inverted input
F9	F9	Coil
F11	F11	Reverse coil
sF3	Shift + F3	Set (latch) coil
sF4	Shift + F4	Reset (unlatch) coil
sF5	Shift + F5	Positive conversion detection coil
sF6	Shift + F6	Negative conversion detection coil
F10	F10	Function / Function block
sF7	Shift + F7	Extended function
c3	Ctrl + 3	Normally open OP contact
c4	Ctrl + 4	Normally closed OP contact
c5	Ctrl + 5	Positive conversion detection OR contact
c6	Ctrl + 6	Negative conversion detection OR contact

* In shortcut key expressions for the editing toolbar, 's' represents the Shift key, and 'c' represents the Ctrl key.

Example: Positive conversion detection contact: Shift + F1 → s + F1 → sF1

3. Contact Input

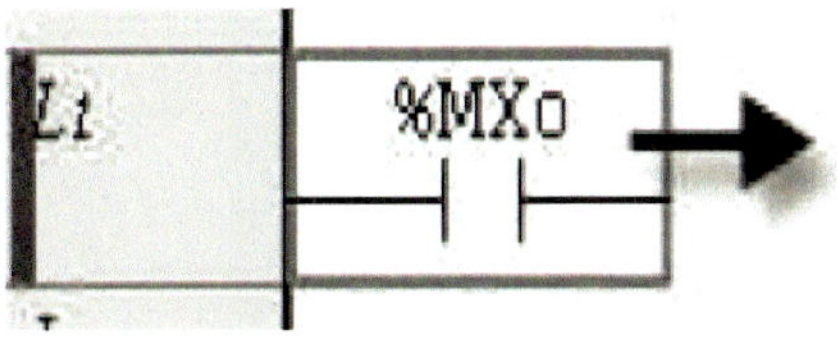

① Move the cursor to the position where you want to input the contact.

② Select the type of contact to be input from the toolbar and click the editing area, or press the corresponding shortcut key for the contact you want to input.

③ In the variable input dialog box, enter the device name and click OK. For detailed information about the variable input dialog box, refer to "Variable/Device Input."

④ Select [Tools] - [Options] from the menu. If the "Immediate Input Mode" is disabled on the 'Edit' page, the variable input dialog box will not automatically appear when inserting contacts or coils.

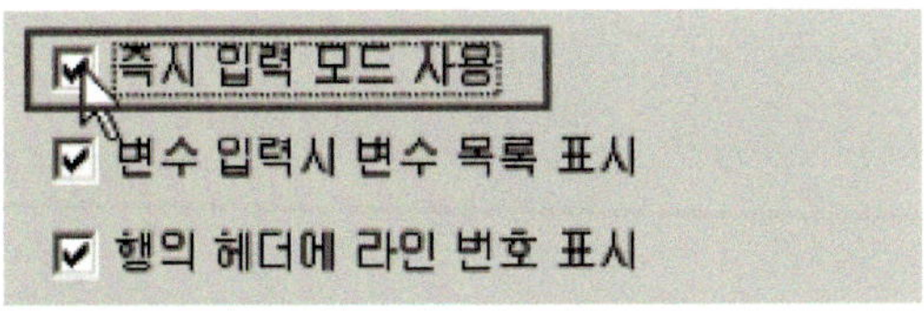

⑤ During editing, the default behavior of the Enter key is to input the same type of contact (coil) as the one previously entered during the last edit.

4. Variable/Device Input

After moving the cursor to the position where you want to input the variable, select [Edit] - [Select/Add Variable] from the menu.

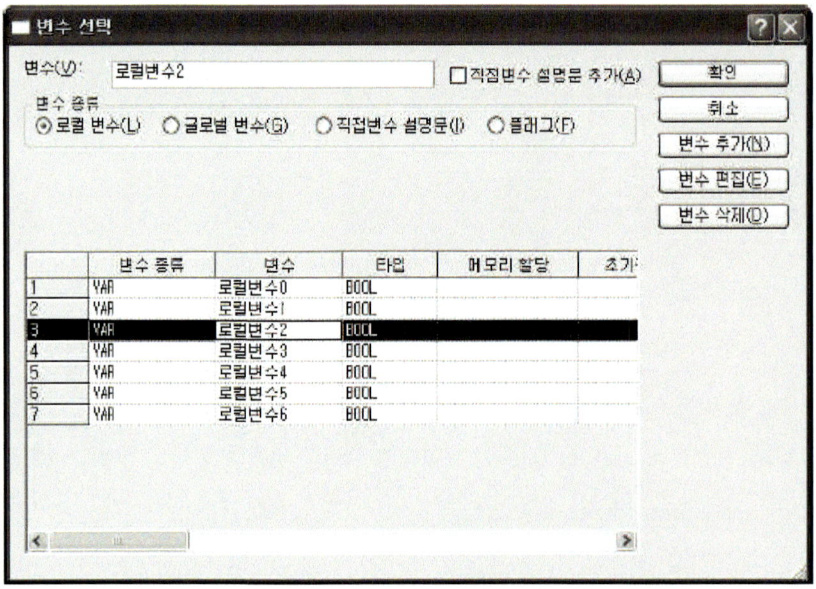

5. Variable/Device Input

ⓐ Variable (V): You can input a constant, direct variable, or declared variable name. If the entered string is not registered as a variable in the local variable list, the "Add Variable" dialog box will appear.

ⓑ Local Variable: Displays the list of declared local variables.

ⓒ Add Variable: Opens the dialog box that allows you to add a variable to the local variable list.

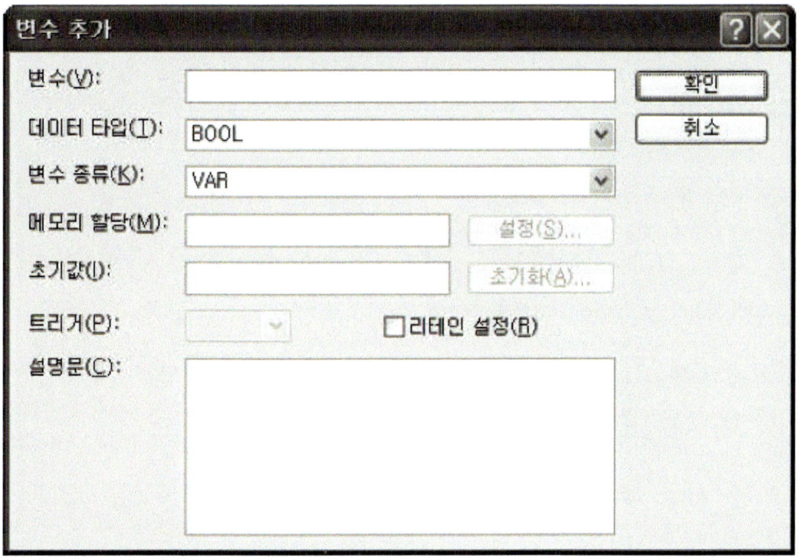

ⓓ Edit Variable: Opens a dialog box that allows you to edit the selected variable.

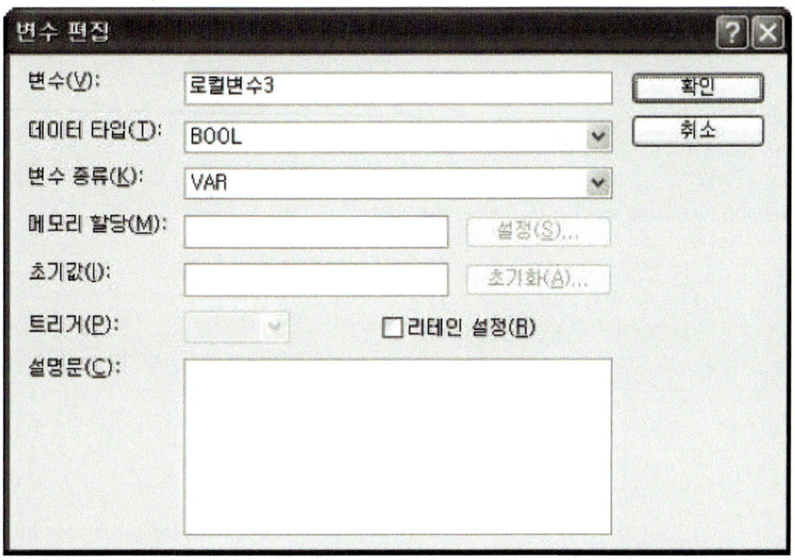

ⓔ Delete Variable: Deletes the selected variable from the local variable list.

ⓕ OK: Applies the entered or selected items and closes the dialog box.

ⓖ Cancel: Closes the dialog box without applying changes.

Sequence Program Rendering

1. Contacts and Coils

(1) -| |- Normally Open Contact (a contact) - F3

- The logical state of the specified bit address is transferred from left to right.

- It turns ON when voltage is applied and turns OFF when voltage is not applied.

(2) -|/|- Normally Closed Contact (b contact) - F4

- The logical state of the specified bit address is inverted and transferred from left to right.

- It turns OFF when voltage is applied and turns ON when voltage is not applied.

(3) -( )- Output Coil - F9

The result of the logical operation of the condition contacts is stored at the specified bit address.

(4) -|P|- Positive Conversion Detection Contact - Shift + F1

When the logical value of the specified bit address changes from OFF to ON, it stays ON for 1 scan time.

(5) -|N|- Negative Conversion Detection Contact - Shift + F2

When the logical value of the specified bit address changes from ON to OFF, it stays ON for 1 scan time.

(6) Function Block: Used when setting blocks with special functions.

2. Direct variable

When the %IX0.0.0 switch is turned ON, the %QX0.1.0 LED lamp turns ON, and due to self-holding, even if the %IX0.0.0 switch is turned OFF, the %QX0.1.0 LED lamp does not turn OFF. When the %IX0.0.1 switch is turned ON, the %QX0.1.0 LED lamp turns OFF.

(1) Time Chart

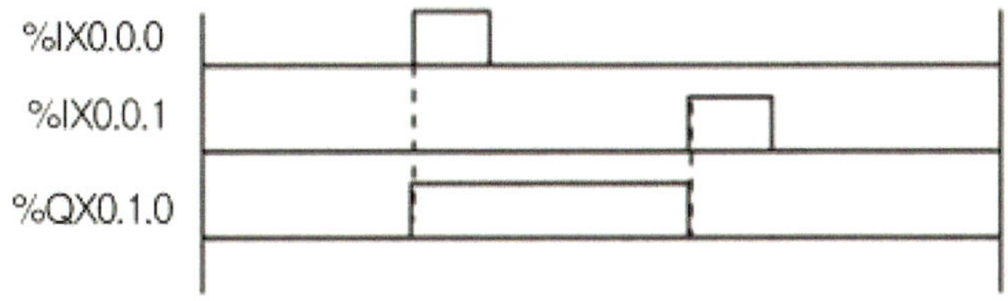

(2) Program

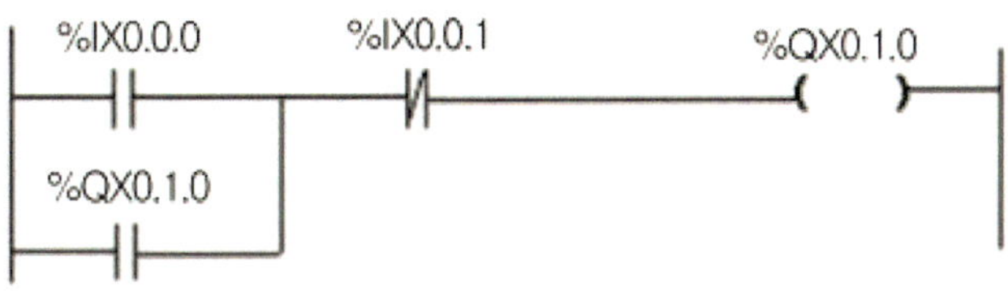

3. Creating a Project

(1) Click the Start Menu () in Windows, select Programs > XG5000, or click the XG5000 shortcut icon () on the desktop. The XG5000 initial screen will appear as shown below.

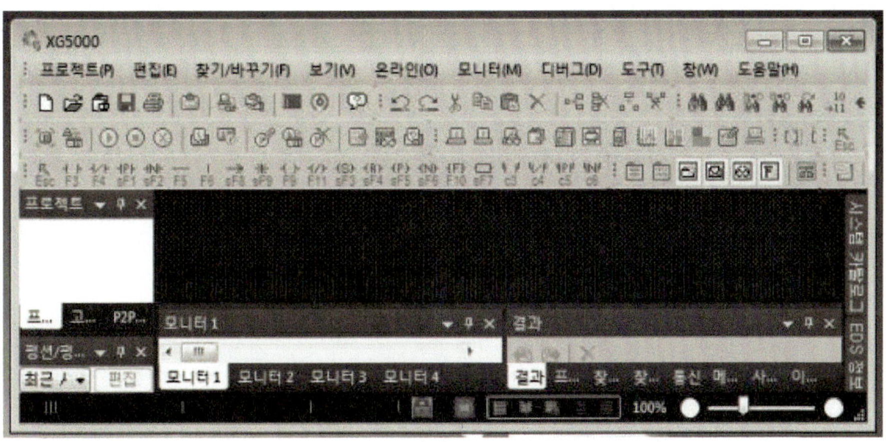

(2) If you saved a project before exiting XG5000, the last saved project will appear on the screen. When you click the project menu bar, a pop-up menu, as shown in the figure below, will appear.

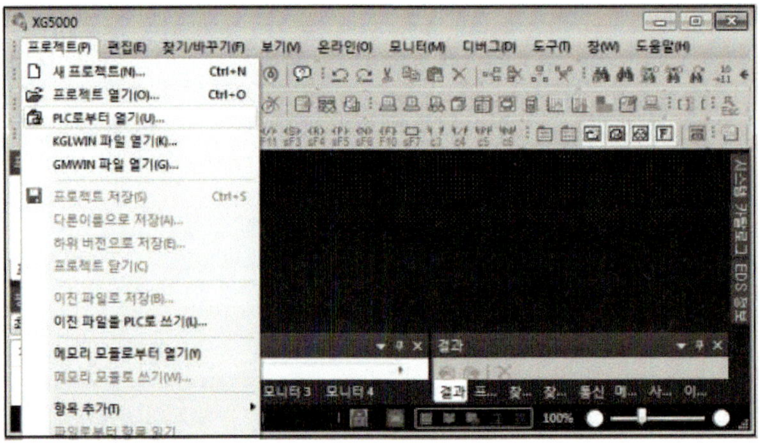

4. Program Editing

(1) To edit a normally open contact (a contact) in the program, click the normally open contact icon (┤├) in the toolbox, and then click again in the program window. In the variable selection window, enter the variable %IX0.0.0 and click "OK(확인)."

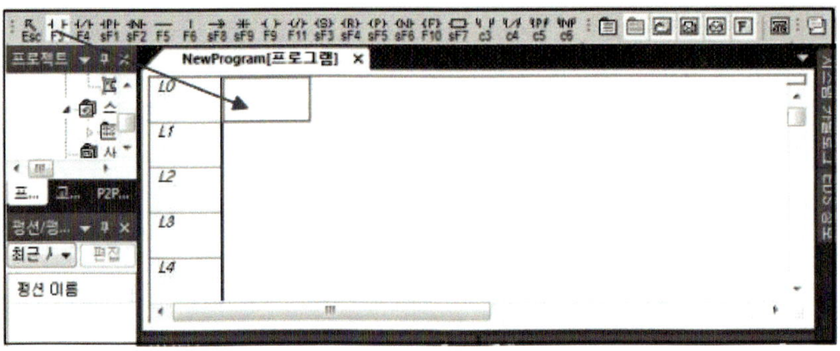

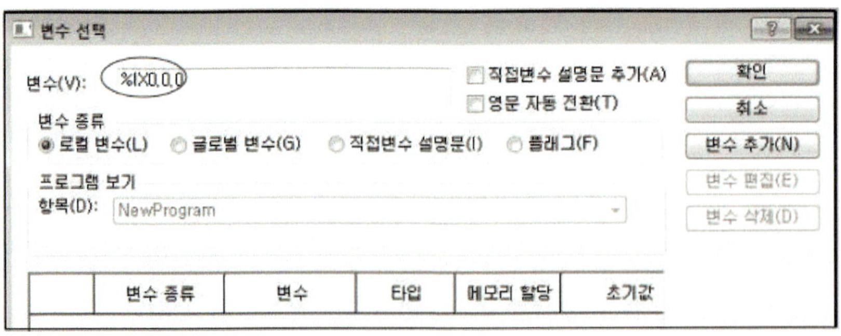

(2) To edit a normally closed contact (b contact) in the program, click the normally closed contact icon ( ) in the toolbox, and then click again in the program window. In the variable selection window, enter the variable %IX0.0.1 and click "OK(확인)."“

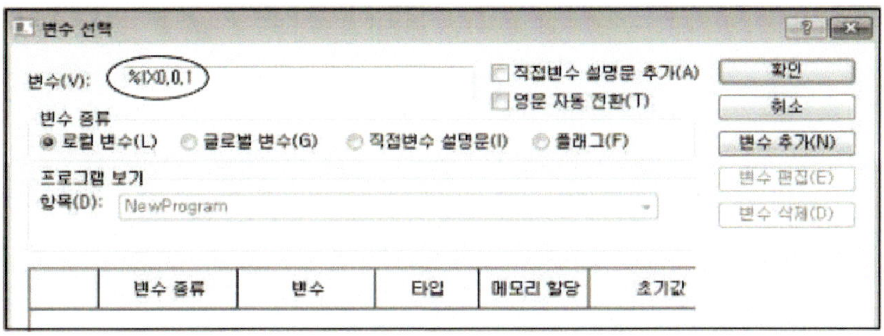

(3) To edit an output coil in the program, click the output coil icon ( ) in the toolbox, and then click again in the program window. In the variable selection window, enter the variable %QX0.1.0 and click "OK(확인)."

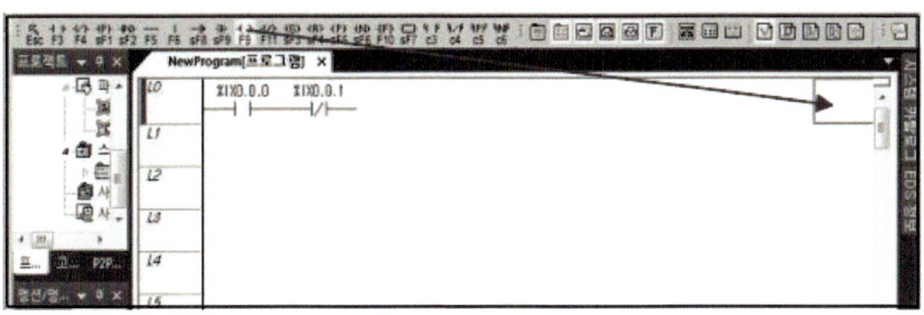

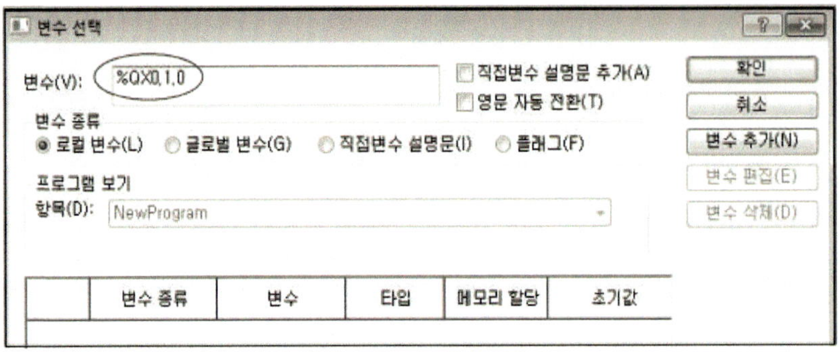

(4) To edit a vertical line in the program, click the vertical line icon ( ) in the toolbox, and then click again in the program window.

(5) When you click in the program window, the variable selection window will appear. If you want to add a description to the address corresponding to the variable name, click "Add Direct Variable Description(직접 변수 설명문 추가)" and then click "OK(확인).“

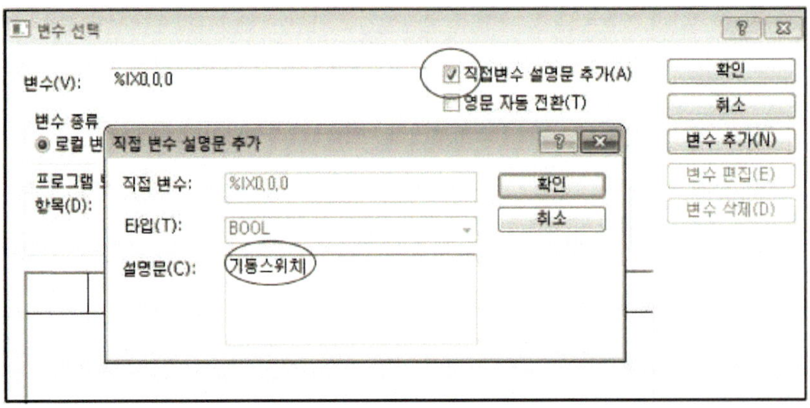

(6) When you click "Add Direct Variable Description," a window like the one shown in the figure will appear. Enter the description for the variable as "기동스위치" (you can create this freely, for example, "start_sw"), and then click "OK(확인)."

(7) Enter "%IX0.0.1" as the variable name, then click "Add Direct Variable Description" and enter "정지스위치" (you can create this freely, for example, "stop_sw") as the description. After entering "QIX0.1.0" as the variable name, click "Add Direct Variable Description" and enter "램프 (Lamp, you can create this freely, for example, " L ")" as the description.

(8) Once the address assignment for all devices is complete, select "보기(View)" menu >> "디바이스/설명문 보기(Show Device/Description)" to view the descriptions assigned to each variable in the program you've created.

(9) Inserting Descriptions

You can insert a description related to the operation of specific parts of the program. Double-click the line number where you want to insert the description, and a "설명문/레이블 (Description/Label)" screen like the one below will appear.

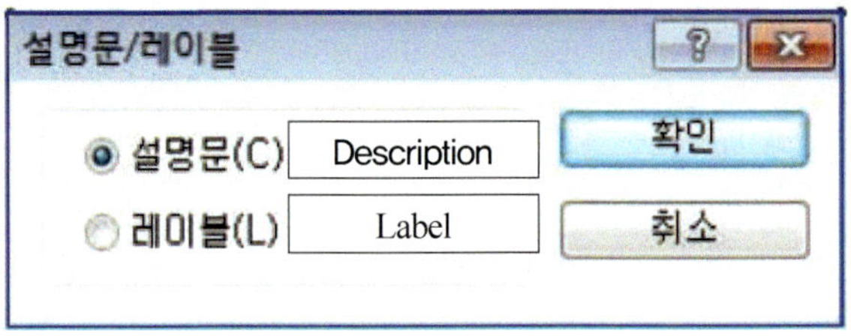

After selecting "Description" and clicking "확인(OK)," a description input screen like the one shown below will appear. Enter the desired description and click "확인(OK)."

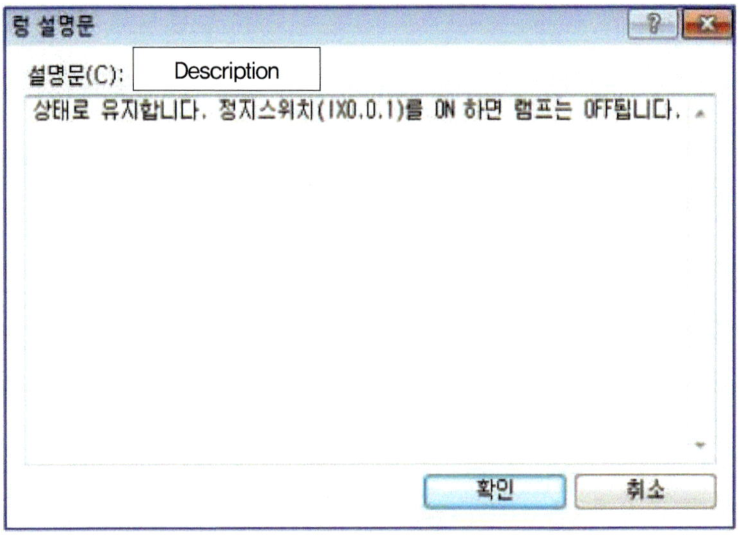

(10) Verifying Operation through Simulation

There are two ways to start the operation: by clicking the icon ( ) or by selecting 메뉴(Menu) >> 도구(T) (Tools) >> 시뮬레이터 시작(s) (Start Simulator). When the simulation starts, the following screens will appear in sequence. Click "OK" in the "쓰기(Write)" window to begin the operation.

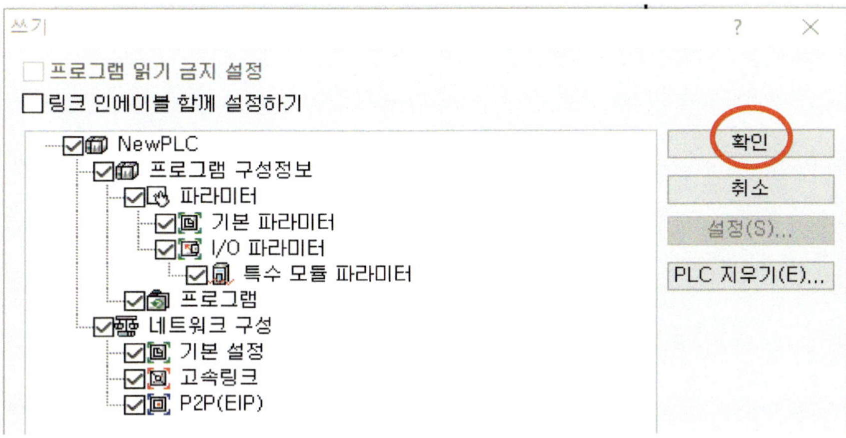

When the screen below appears, click "예(Y).“

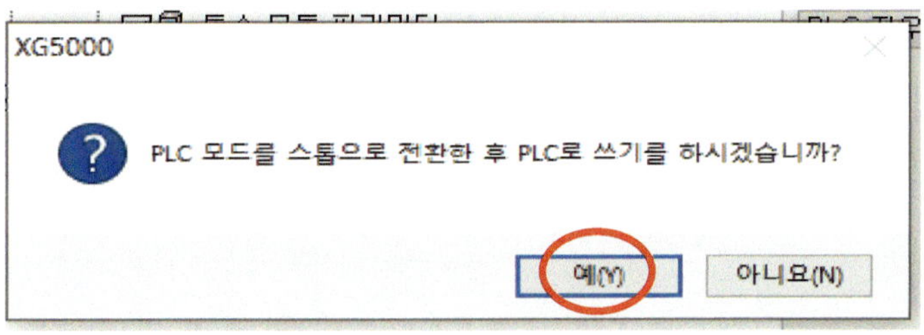

When the simulation starts, a screen like the one below will appear.

To control the operation of the contacts, as shown in the image below, click the desired input contact and right-click with the mouse. A menu will appear, and you can select "현재값 변경...(Change Current Value)"

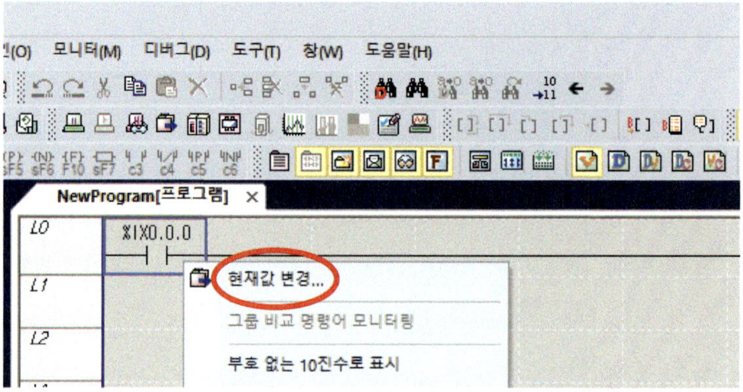

When a window titled "현재 값 변경(Change Current Value)" appears, select "ON" or "OFF" under "현재 값 입력(Enter Current Value)" and click to apply the change.

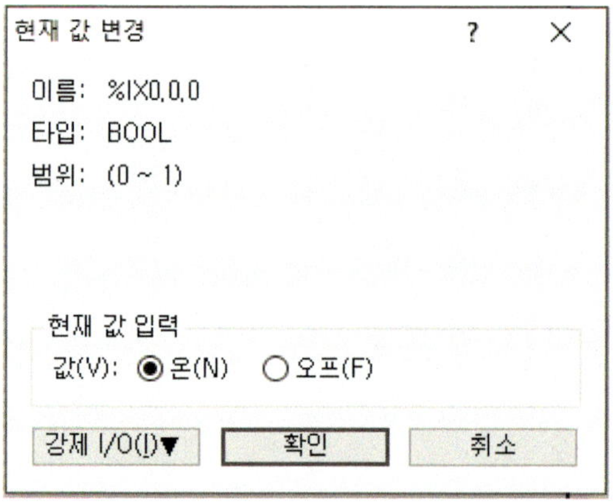

When the a contact of "%IX0.0.0" is turned ON (N) as shown below, it connects in blue, and the output "%QX0.0.0" is activated, which is also indicated in blue.

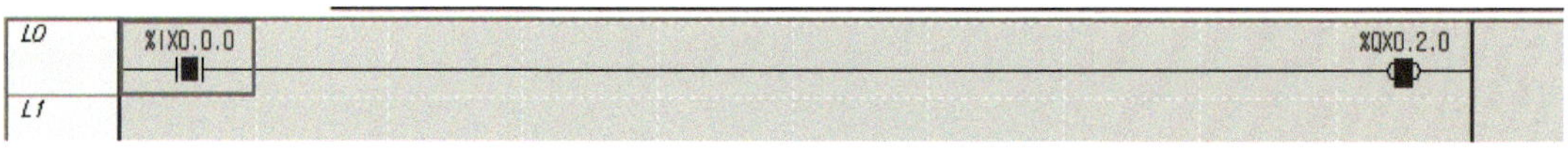

2. PLC-XGT Job Training Practice Example

Practice Example 1 ; AND

❑ Conditions

- The relay coil R is energized and the R contact closes to light the lamp only when both inputs A and B are ON, forming an AND circuit.

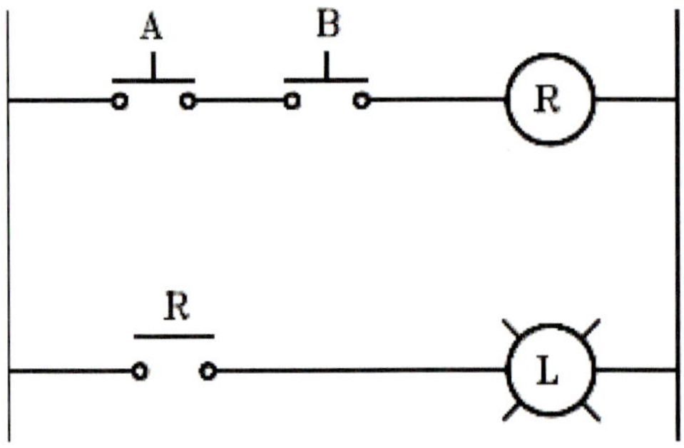

Relay Circuit

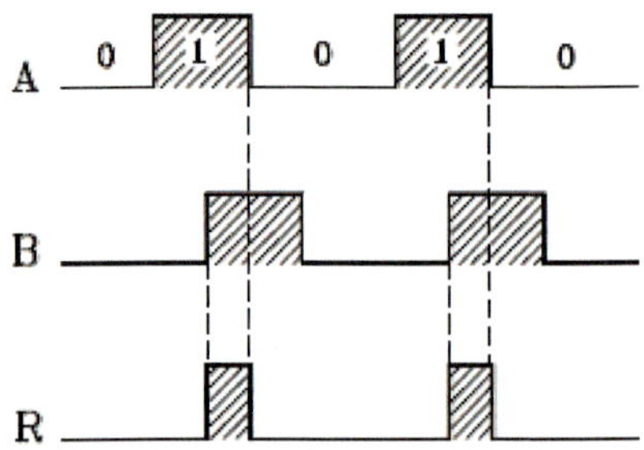

Time Chart

❑ PLC rendering

Practice Example 2 ; OR

❑ Conditions

- An OR circuit is a circuit that produces an output when one or more of several input signals are ON and is referred to as a parallel circuit.

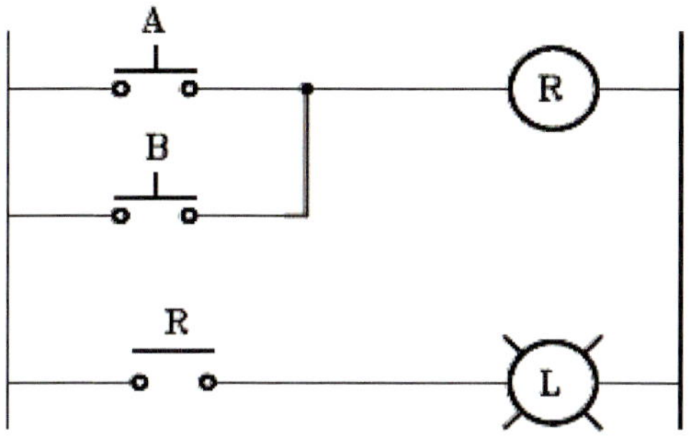

Relay Circuit

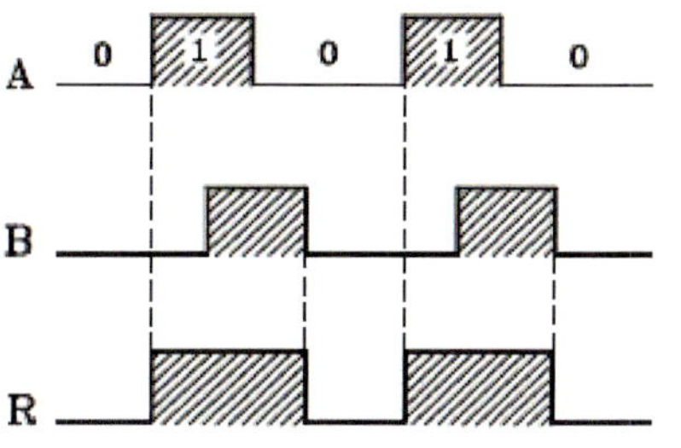

Time Chart

❑ PLC rendering

Practice Example 3 ; Self-holding circuit

❑ Conditions

- The operating principle is as follows: when the push-button switch PB1 is pressed, the relay is activated, and both contacts R(1) and R(2) close simultaneously, causing the lamp to light up. Even after releasing the push-button switch PB1, the current continues to flow through contact R(1) and PB2 to the coil, maintaining the operation. In other words, even after PB1 returns to its original state, the relay's operation circuit is sustained by contact R(1). The self-holding is released when the push-button switch PB2 is pressed, causing the relay to return to its original state, opening contacts R(1) and R(2), and resetting the circuit to its initial condition.

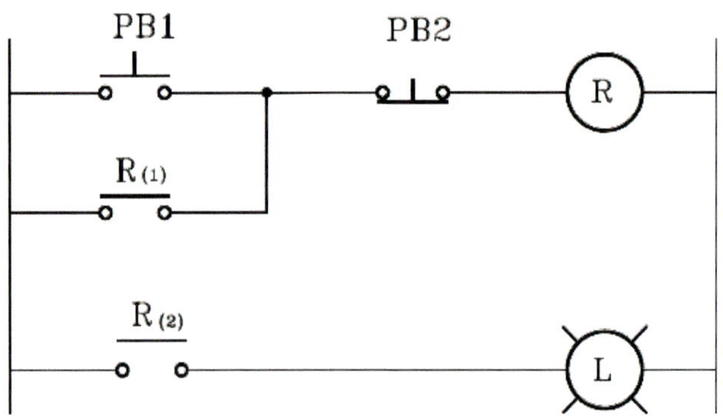

Relay Circuit

Practice Example 4 ; Inter-lock circuit

❑ Conditions

- An interlock circuit is a circuit that uses contacts indicating the operating status of a device to prevent the operation of related equipment, ensuring the protection of the device and the safety of the operator. It is also referred to as a "preceding operation priority circuit" or a "mutual operation prohibition circuit."

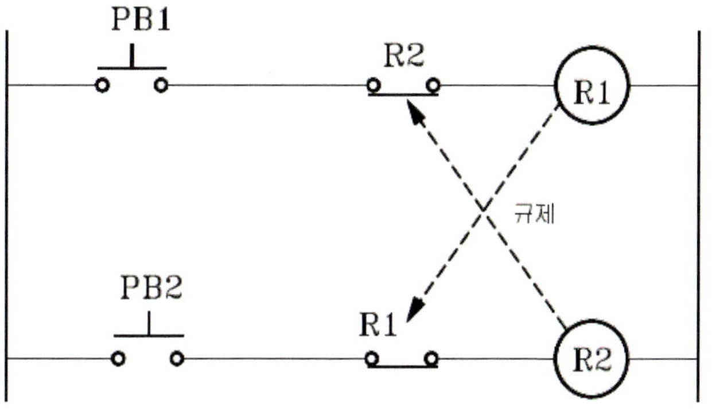

Relay Circuit

Practice Example 5 ; Timer Circuit

❑ Conditions

Function	Description
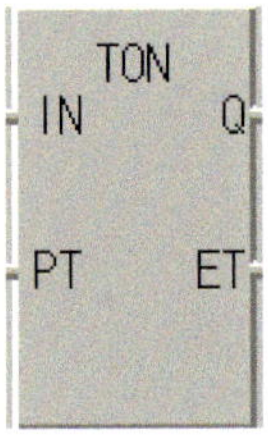 	Input IN : starting conditions for a timer PT : Preset Time Output Q : output of a timer ET : Elapsed Time

Function	Description
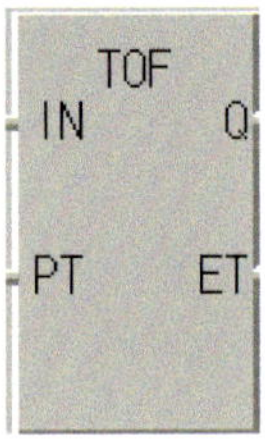 	Input IN : starting conditions for a timer PT : Preset Time Output Q : output of a timer ET : Elapsed Time

❑ PLC rendering

Practice Example 5 ; ON-delay

❑ Conditions

- A delay circuit is a circuit designed so that the output does not appear immediately after receiving an input signal but instead appears after a planned delay. There are two types of delay circuits: an ON-time delay circuit and an OFF-time delay circuit.

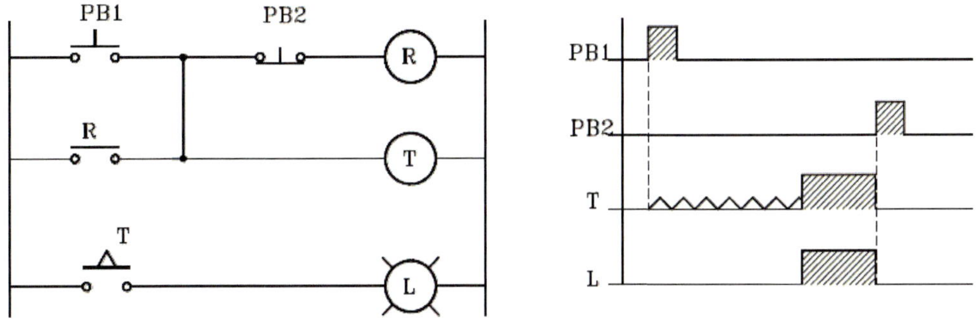

Relay Circuit

Practice Example 6 ; OFF-time delay

❑ Conditions

- An off-delay circuit is a circuit in which, when a return signal is given, the output does not return immediately, but the load is opened after a planned delay. The circuit can be configured using the b-contact of an ON-delay timer or the a-contact of an OFF-delay timer.

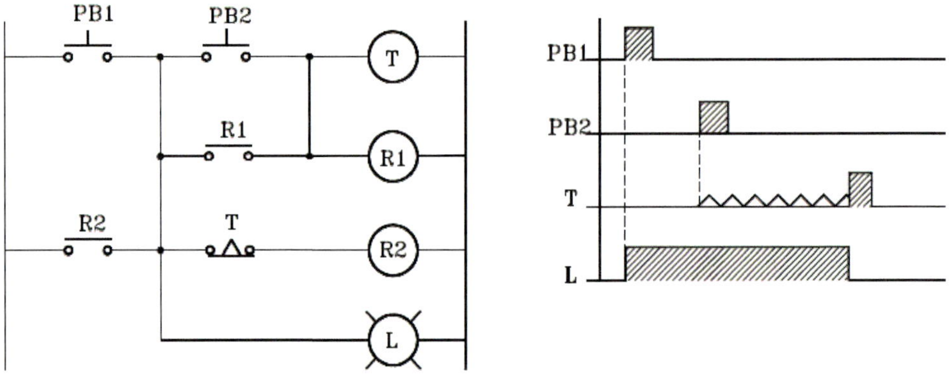

Relay Circuit

Practice Example 7 ; Basic timer circuit

❑ PLC rendering 1.

	변수 명	데이터 타입	메모리 할당	초기 값	변수 종류	사용 여부	설명문
1	L0	BOOL	%QX0.2.0		VAR		
2	PB0	BOOL	%IX0.0.0		VAR		
3	PB1	BOOL	%IX0.0.1		VAR		
4	R0	BOOL	<자동>		VAR		
5	T0	FB Instance	<자동>		VAR		

❑ PLC rendering 2.

Practice Example 8 ; Automatic stop circuit

❑ PLC rendering

Practice Example 9 ; Control 2 lamps

❑ Conditions

PB1(start SW.) on → T0 → L0 ON → T1 → L0 OFF / L1 ON

❑ PLC rendering

Practice Example 10 ; Control 3 lamps

❑ PLC rendering

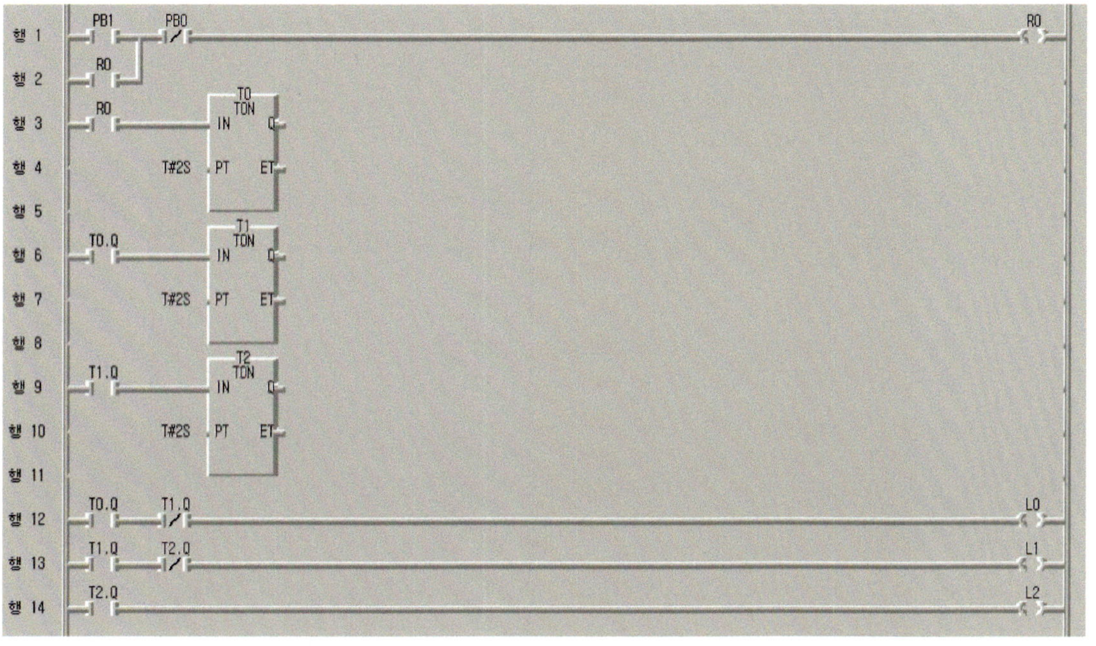

Practice Example 11 ; Flicker circuit

❑ PLC rendering

	변수 명	변수 값	데이터 타입	메모리 할당	초기 값	변수 종류	사용 여부	설명문
1	기동SW	0	BOOL	%IX0.0.0		VAR	*	
2	램프	0	BOOL	%QX0.2.0		VAR	*	
3	정지SW	0	BOOL	%IX0.0.1		VAR	*	
4	R0	1	BOOL	<자동>		VAR	*	
5	T0		FB Instance	<자동>		VAR	*	
6	T1		FB Instance	<자동>		VAR	*	

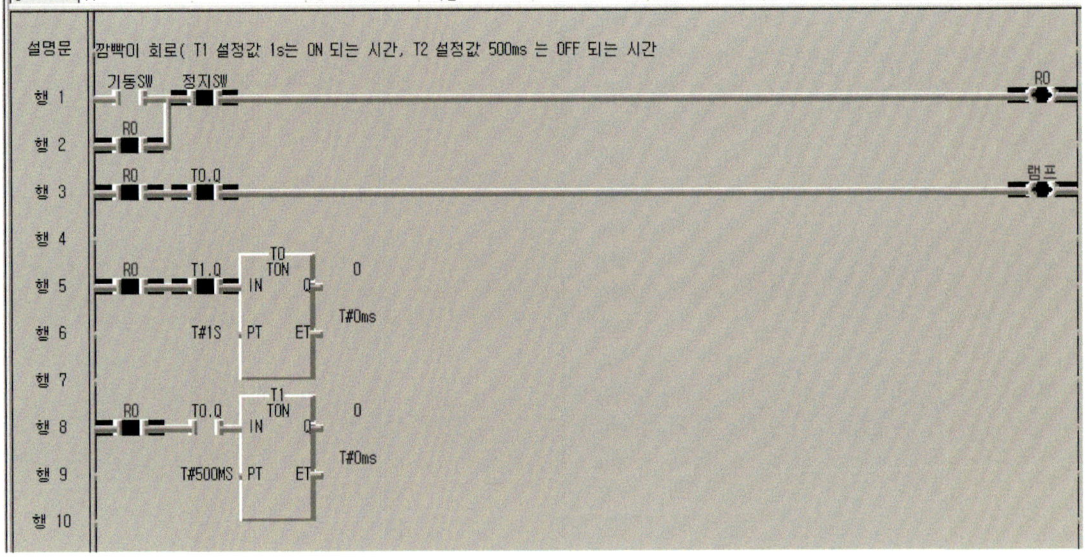

※ Command _T(seconds)S ; a command for inputting a blinking signal every (seconds).

Practice Example 12 ; Automatic valve control for urinal

❑ PLC rendering

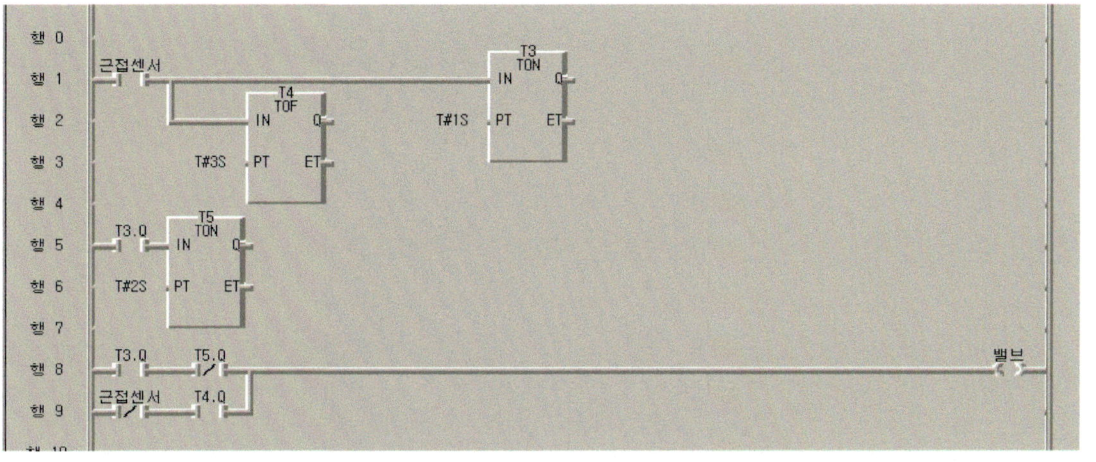

Practice Example 13 ; Repetitive operation circuit

❑ PLC rendering

Practice Example 14 ; Sequential operation circuit for 3 lamps

❑ PLC rendering

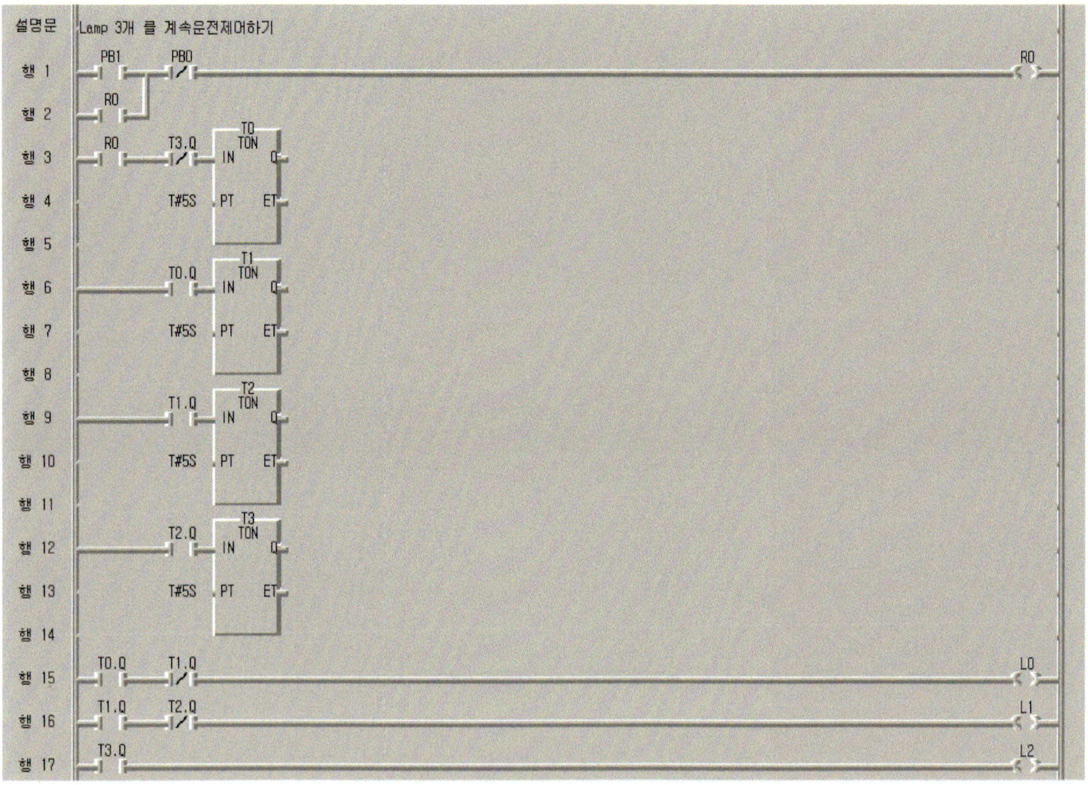

Practice Example 15 ; Motor sequential control

❑ Conditions

(1) Pressing PB0 the first time turns Motor 1 ON.

(2) Pressing PB0 the second time turns Motor 2 ON.

(3) Pressing PB0 the third time turns Motor 3 ON.

(4) Pressing PB1 stops the operation of all motors.

❑ PLC rendering

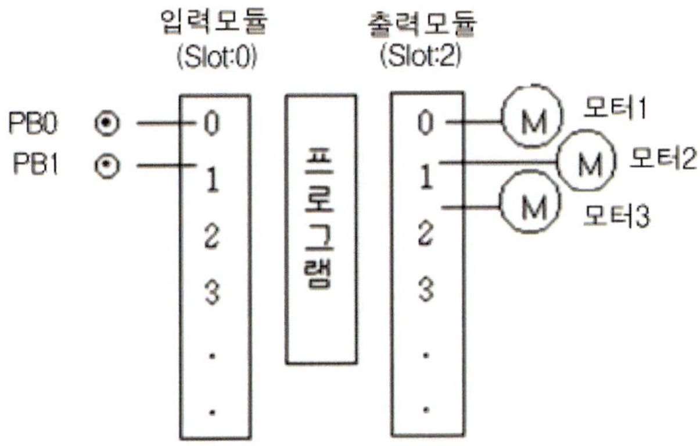

❑ Time Chart

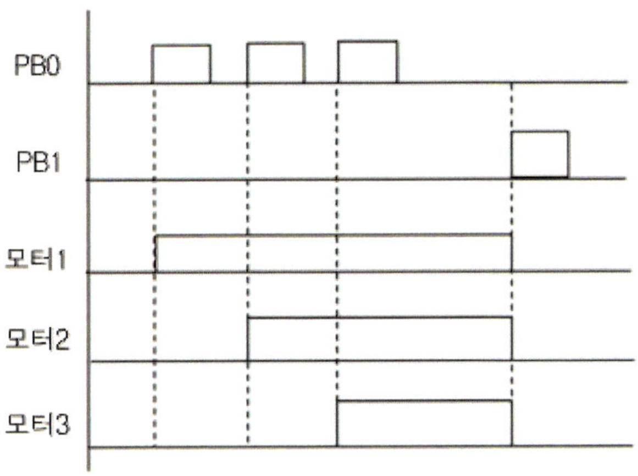

❑ PLC rendering

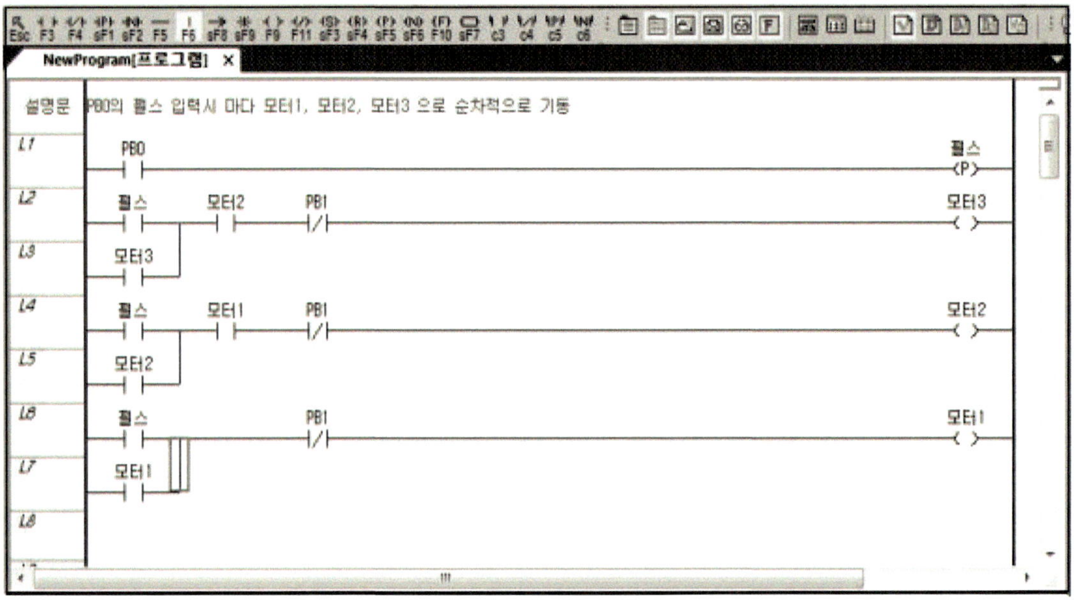

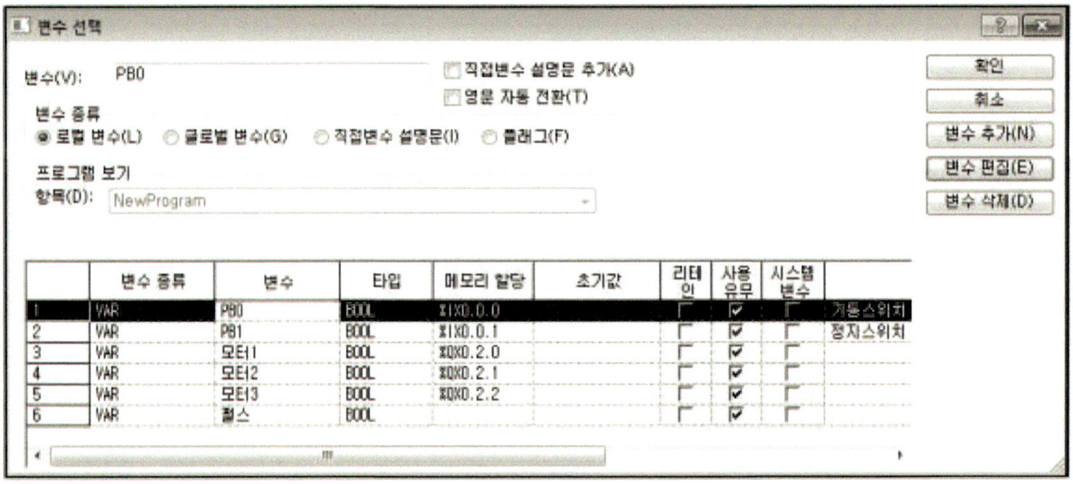

PART II

MITSUBISHI PLC (GX-Works2)

1.
GX-Works2

PLC I/O Address Allocation

1. Check the number of I/O points used by the input/output modules and intelligent function modules installed in the base slot.

(1) Input and output use 16, 32, or 64 points depending on the module type.

(2) Among intelligent modules, A/D and D/A use 16 points each for input and output, while other modules use 32 points each for input and output.

(3) An empty slot uses 16 points.

(4) I/O addresses use a hexadecimal system.

(5) Input addresses begin with "X" and output addresses begin with "Y."

ex) X00 ~ X1F / Y00 ~ Y1F

2. Types of Memory

(1) Input (X): Memory that stores the ON/OFF status of switches and sensors connected to the PLC input module.

(2) Output (Y): Memory that stores the ON/OFF information of devices like lamps and relays connected to the PLC output.

(3) Internal Bit (M): Memory that stores the ON/OFF information of devices like lamps and relays

connected to the PLC output but does not occupy an output address.

ex) Y20 → M00

(4) Data Register (D):

- Memory used to store word-sized (16-bit) information necessary for writing PLC programs.
- Can also be used by dividing the 16-bit memory into individual bits as needed.
- To use it as bits, the format "word address.bit address" is applied.

Execution Order of a PLC Program

The program executes from top to bottom and from left to right, similar to the XGT program.

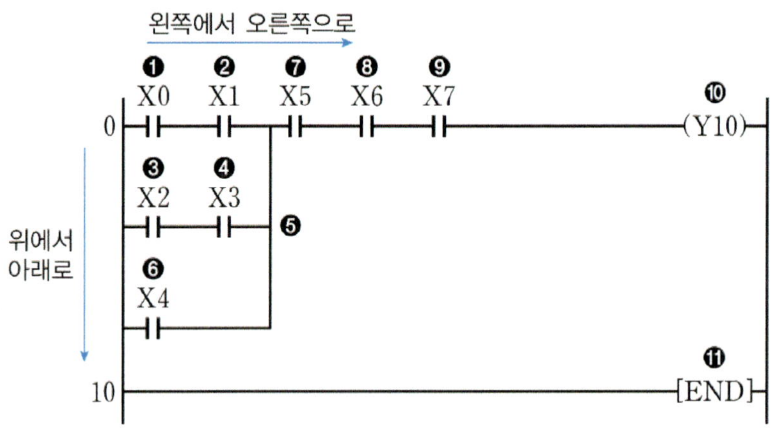

Screen Configuration of GX-Works2

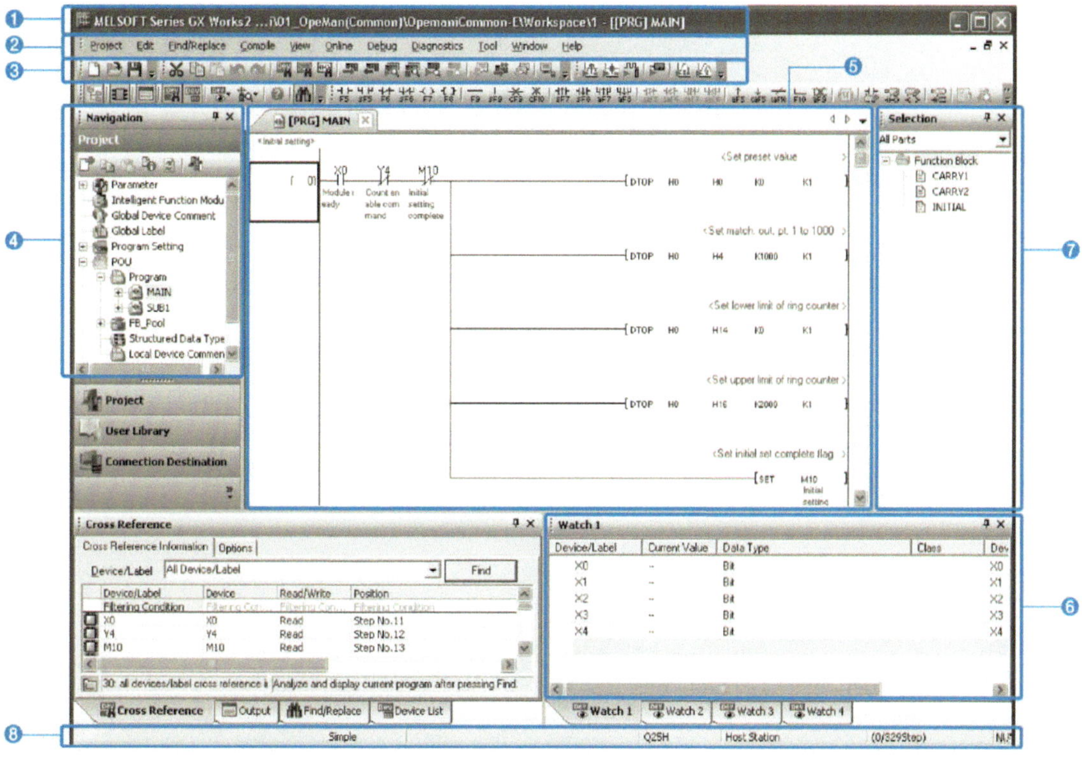

① Title Bar

② Menu Bar

③ Toolbar

④ Navigation Window

⑤ Work Window

⑥ Connection Window

⑦ Component Selection Window

⑧ Status Bar

Main Shortcut Keys

Shortcut Keys	Icons	Functions
F2		Write Mode
F4	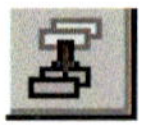	Compile
F5		a Contacts
F6	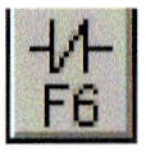	b Contacts
Shift + F5		a Contacts OR
Shift + F5		b Contacts OR
F6		Coil
F8	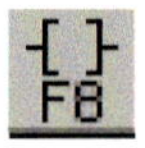	Function (Applied Instruction)
Shift + F9		Vertical Line
F9		Horizontal Line
Shift + F7		Pulse Rising
Shift + F8		Pulse Falling
Alt + F7		Pulse Rising OR
Alt + F8		Pulse Falling OR

2. MITSUBISHI PLC Job Training Practice Example

Practice Example 1 – Lamp Lighting Circuit

1. Practical Task : Lamp Lighting Circuit

2. Operating Condition

The lamp turns on when the push-button switch (PBS) is pressed (ON).)

3. Address and Device Description

	Name	Function	I/O Allocation
Input	PBS	푸쉬 버튼스위치 Push Button Switch	X00
Output	Lamp	램프Lamp	Y20

❒ PLC rendering

Practice Example 2 - Lighting Circuit

1. Practical Task

The lamp is turned on using the PBS button.

2. Operating Condition

When the PBS is ON, the lamp lights up.

3. Address and Device Description

	Name	Function	I/O Allocation
Input	PBS	Push Button Switch	X00
Output	Lamp	Lamp	Y20

❐ PLC rendering

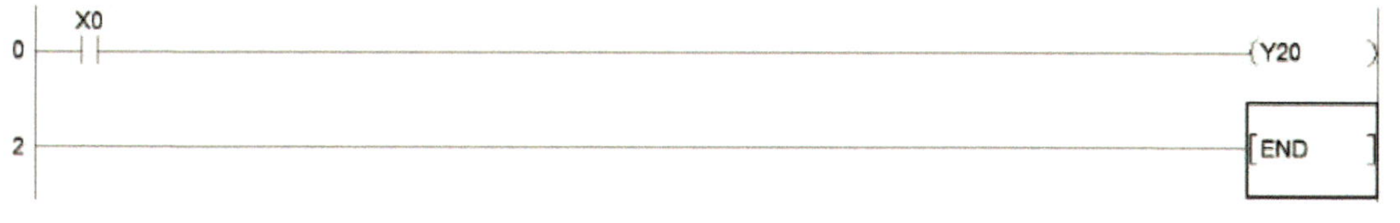

Practice Example 3 - (AND Circuit)

1. Practical Task

Understanding the AND Circuit and Circuit Rendering

2. Operating Condition

The lamp lights up only when both PBS1 and PBS2 are ON simultaneously.

3. Address and Device Description

	Name	Function	I/O Allocation
Input	PBS 1	Push Button Switch	X00
	PBS 2	Push Button Switch	X01
Output	Lamp	Lamp	Y20

❒ PLC rendering

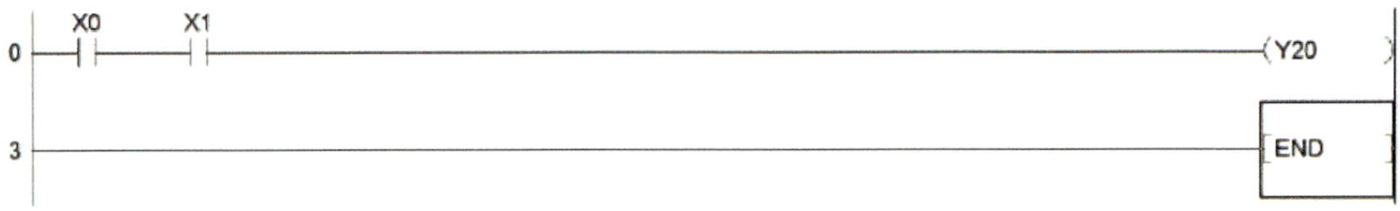

Practice Example 4 - OR Circuit

1. Practical Task

Understanding the OR Circuit and Circuit Rendering

2. Operating Condition

The lamp lights up when either PBS1 or PBS2, or both, are ON.

3. Address and Device Description

	Name	Function	I/O Allocation
Input	PBS 1	Push Button Switch	X00
	PBS 2	Push Button Switch	X01
Output	Lamp	Lamp	Y20

❐ PLC rendering

Practice Example 5 - Stop-Priority Self-Holding Circuit

1. Practical Task

Understand the self-holding circuit and render the circuit.

2. Operating Condition

- When PBS1 is ON, the lamp lights up.
- Even if PBS1 is turned OFF, the lamp remains lit.
- When PBS2 is ON, the lamp turns off.

3. Address and Device Description

	Name	Function	I/O Allocation
Input	PBS 1	Push Button Switch	X00
	PBS 2	Push Button Switch	X01
Output	Lamp	Lamp	Y20

❒ PLC rendering

Practice Example 6 - Operation-Priority Self-Holding Circuit

1. Practical Task

Understanding the Priority Self-Holding Circuit and Circuit Rendering.

2. Operating Condition

- When PBS1 is ON, the lamp lights up.
- Even if PBS1 is turned OFF, the lamp remains lit.
- When PBS2 is ON, the lamp turns off.
- If both PBS1 and PBS2 are ON simultaneously, the lamp lights up.

3. Address and Device Description

	Name	Function	I/O Allocation
Input	PBS 1	Push Button Switch	X00
	PBS 2	Push Button Switch	X01
Output	Lamp	Lamp	Y20

❒ PLC rendering

Practice Example 7 - interlock circuit

1. Practical Task

Understand the interlock circuit (mutual exclusion circuit) and render the circuit.

2. Operating Condition

- When the circuit operation switch PBS1 is ON, the internal allocated memory M0 is energized.
- When the circuit operation switch PBS2 is ON, the internal allocated memory M1 is energized.
- When the circuit operation switch PBS3 is ON, the internal allocated memory M2 is energized.
- If M1 is energized while M0 is energized, Lamp 1 lights up.
- If M2 is energized while M0 is energized, Lamp 2 lights up.
- PBS1 and PBS2 are configured as an interlock circuit to form a mutual exclusion circuit.

3. Address and Device Description

	Name	Function	(I/O Allocation)
Input	PBS1	Push Button Switch	X00
	PBS2	Push Button Switch	X01
	PBS3	Push Button Switch	X02
Output	Lamp1	Lamp	Y20
	Lamp2	Lamp	Y21

❒ PLC rendering

Practice Example 8 - Positive/negative Conversion Detection Circuit

1. Practical Task

Understand the positive/negative conversion detection circuit and render the circuit.

2. Operating Condition

- When PBS1 is ON, a single pulse is triggered momentarily, lighting up Lamp 1 (positive conversion).
- When PBS2 changes from ON to OFF, a single pulse is triggered momentarily, lighting up Lamp 2 (negative conversion).
- When PBS3 is ON, both Lamp 1 and Lamp 2 turn off.

3. Address and Device Description

	Name	Function	I/O Allocation
Input	PBS1	Push Button Switch	X00
	PBS2	Push Button Switch	X01
Output	Lamp1	Lamp	Y20
	Lamp2	Lamp	Y21

❒ PLC rendering

Practice Example 9 - SET/RESET Circuit

1. Practical Task

Understand the SET/RESET circuit and render the circuit.

2. Operating Condition

- When PBS1 is ON, Lamp 1 lights up (SET).
- When PBS2 is ON, Lamp 1 turns off (RESET).

3. Address and Device Description

	Name	Function	I/O Allocation
Input	PBS1	Push Button Switch	X00
	PBS2	Push Button Switch	X01
Output	Lamp1	Lamp	Y20

❐ PLC rendering

Practice Example 10 - Timer circuit

1. Practical Task

Understand the timer circuit and render the circuit.

2. Operating Condition

- When PBS1 is ON, the internal allocated memory M0 is energized.
- M0 holds itself (self-latching) after PBS1 is pressed.
- When M0 is energized, Timer (T0) starts a 5-second delay.
- After 5 seconds, T0 activates the Lamp.
- When PBS2 is ON, M0 is de-energized, turning off the Lamp.

3. Address and Device Description

	Name	Function	I/O Allocation
Input	PBS1	Push Button Switch	X00
	PBS2	Push Button Switch	X01
Output	Lamp1	Lamp	Y20

❐ PLC rendering

Practice Example 11 - Timer circuit-2

1. Practical Task

Understand how to use a timer circuit to turn off a lamp and render the circuit.

2. Operating Condition

- When PBS1 is ON, the lamp lights up.
- After 3 seconds, the lamp automatically turns off.

3. Address and Device Description

	Name	Function	I/O Allocation
Input	PBS1	Push Button Switch	X00
Output	Lamp1	Lamp	Y20

❐ PLC rendering

Practice Example 12 - Flicker circuit

1. Practical Task

Understand the flicker circuit and render the circuit.

2. Operating Condition

- When PBS1 is ON, Lamp1 lights up. After 3 seconds, Lamp1 turns off, and Lamp2 lights up, repeating this process infinitely.
- Two timers, T1 and T2, are used with a preset time of 3 seconds each to control the alternating sequence.

3. Address and Device Description

	Name	Function	I/O Allocation
Input	PBS1	Push Button Switch	X00
Output	Lamp1	Lamp	Y20
	Lamp2	Lamp	Y21

❐ PLC rendering

Practice Example 13 - start detection circuit

1. Practical Task

Understand the start detection circuit and render the circuit.

2. Operating Condition

- When PBS1 is ON, the internal allocated memory M0 is energized, and the lamp lights up.
- After 5 seconds, the timer (T0) turns off the lamp.

3. Address and Device Description

	Name	Function	I/O Allocation
Input	PBS1	Push Button Switch	X00
Output	Lamp	Lamp	Y20

❐ PLC rendering

Practice Example 14 - STMR Circuit

1. Practical Task

Understand the STMR circuit and render the circuit.

2. Operating Condition

- When PBS1 is ON, the STMR circuit starts operating (with M1 being activated first).
- When M1 is energized, Lamp1 lights up, When M2 is energized, Lamp2 lights up, When M3 is energized, Lamp3 lights up. When M4 is energized, Lamp4 lights up.

3. Address and Device Description

	Name	Function	I/O Allocation
Input	PBS1	Push Button Switch	X00
Output	Lamp 1	Lamp	Y21
	Lamp 2	Lamp	Y22
	Lamp 3	Lamp	Y23
	Lamp 4	Lamp	Y24

❒ PLC rendering

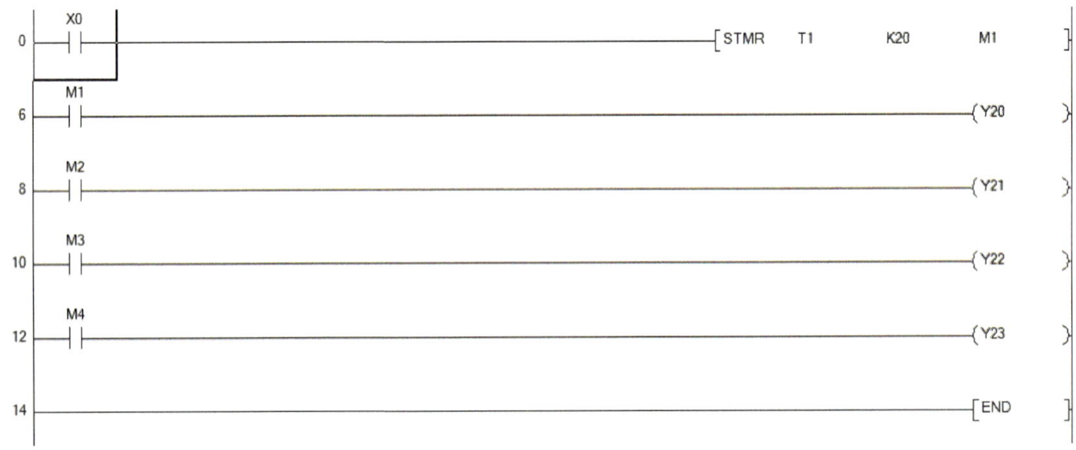

Practice Example 15 - 1-second blinking infinite loop

1. Practical Task

Understanding and rendering a 1-second blinking circuit using a timer

2. Operating Condition

- When PBS1 is ON, the lamp blinks at 1-second intervals.
- When PBS2 is ON, the lamp turns off.

3. Address and Device Description

	Name	Function	I/O Allocation
Input	PBS1	Push Button Switch	X00
Output	Lamp	Lamp	Y20

❐ PLC rendering

Practice Example 16 - An infinite loop circuit for three lamps lighting up

1. Practical Task

Understanding the sequential lighting of three lamps in an infinite loop using timers.

2. Operating Condition

- When PBS1 is ON, timers T0, T1, and T2 sequentially receive signals at 5-second intervals, causing Lamp1, Lamp2, and Lamp3 to light up in sequence, repeating infinitely.
- When PBS2 is ON, the lamps turn off.

3. Address and Device Description

	Name	Function	I/O Allocation
Input	PBS1	Push Button Switch	X01
	PBS2	Push Button Switch	X00
Output	Lamp 1	Lamp	Y20
	Lamp 2	Lamp	Y21
	Lamp 3	Lamp	Y22

❒ PLC rendering

Practice Example 17 - Counter circuit

1. Practical Task

Understanding the counter circuit and rendering the circuit

2. Operating Condition

- When PBS1 is pressed 3 times, the Lamp lights up.
- When PBS2 is pressed, the counter resets.

3. Address and Device Description

	Name	Function	I/O Allocation
Input	PBS1	Push Button Switch	X00
	PBS2	Push Button Switch	X01
Output	Lamp	Lamp	Y20

❐ PLC rendering

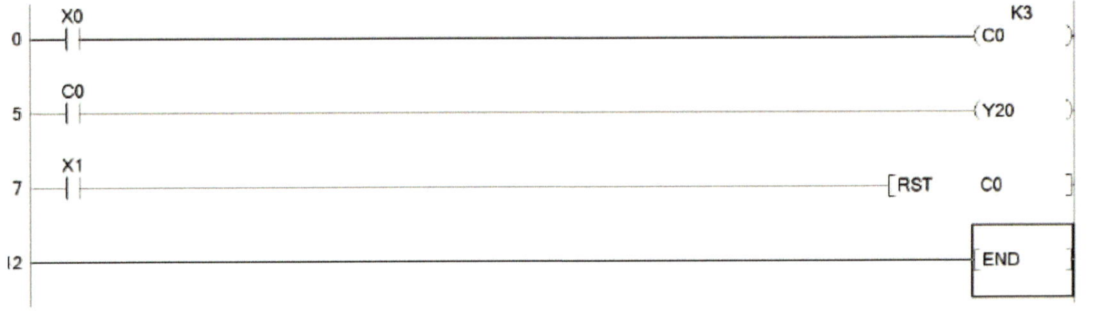

Practice Example 18 - Applied counter circuit

1. Practical Task

Understanding the counter circuit and rendering an applied counter circuit

2. Operating Condition

- When PBS1 is pressed once, Lamp1 lights up.
- When PBS1 is pressed twice, Lamp2 also lights up.
- When PBS1 is pressed three times, both Lamp1 and Lamp2 turn off.

3. Address and Device Description

	Name	Function	I/O Allocation
Input	PBS1	Push Button Switch	X00
	PBS2	Push Button Switch	X01
Output	Lamp 1	Lamp	Y20
	Lamp 2	Lamp	Y21

❒ PLC rendering

Practice Example 19 - Applied counter circuit 2

1. Practical Task

To control the number of times a lamp lights up using a timer and a counter.

2. Operating Condition

- When PBS1 is ON, the internal allocated memory M1 is energized, and M1 holds the PBS1 in a self-latching state.
- The lamp turns ON and OFF at 1-second intervals according to the number of times set in the counter. After reaching the preset count, the system resets.

3. Address and Device Description

	Name	Function	I/O Allocation
Input	PBS1	Push Button Switch	X00
Output	Lamp 1	Lamp	Y20

❒ PLC rendering

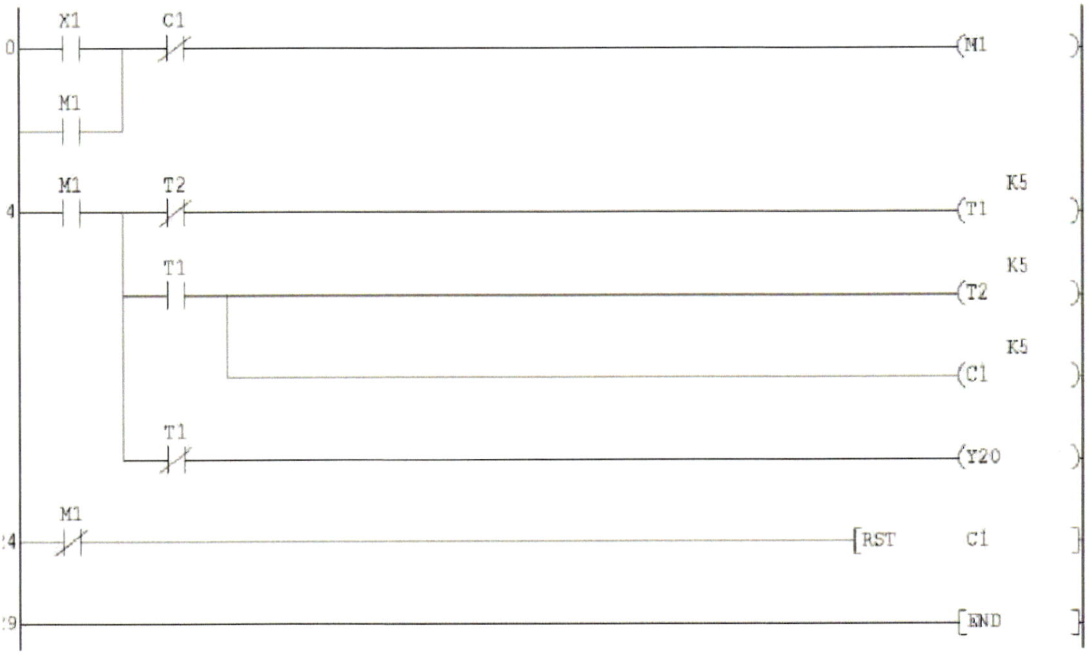

Practice Example 20 - Applied counter circuit 3

1. Practical Task

To control a motor using a counter.

2. Operating Condition

- When PBS1 (positive edge detection input) is ON, the motor (Y20) rotates 5 times, then stops for 5 seconds, and repeats the cycle.

3. Address and Device Description

	Name	Function	I/O Allocation
Input	PBS1	Push Button Switch	X00
Output	Motor	Motor Output	Y20

❐ PLC rendering

Practice Example 21 - MOVE Circuit

1. Practical Task

Understanding the MOVE instruction

2. Operating Condition

- When PBS1 (positive edge detection switch) is ON, Lamp1 lights up. (The MOVE instruction is used, and M0 is applied).
- When PBS2 is ON, Lamp1 turns off.

3. Address and Device Description

	Name	Function	I/O Allocation
Input	PBS1	Push Button Switch	X00
	PBS2	Push Button Switch	X01
Output	Lamp 1	Lamp	Y20

❒ PLC rendering

Practice Example 22 - Timer control circuit using the MOVE instruction

1. Practical Task

Understanding the timer control circuit using the MOVE instruction.

2. Operating Condition

- When PBS1 is ON, Timer T0 is set to 2 seconds using the MOVE instruction.
- When PBS2 is ON, Timer T0 is set to 4 seconds using the MOVE instruction.
- When PBS3 is ON, Lamp1 lights up, and after the timer T0 reaches the set time, Lamp1 turns off.

3. Address and Device Description

	Name	Function	I/O Allocation
Input	PBS1	Push Button Switch	X00
	PBS2	Push Button Switch	X01
	PBS3	Push Button Switch	X02
Output	Lamp 1	Lamp	Y20

❐ PLC rendering

Practice Example 23 - Counter control circuit using the MOVE instruction

1. Practical Task

Understand the counter control circuit using the MOVE instruction.

2. Operating Condition

- When PBS1 is ON, the counter preset is set to 3 using the MOVE instruction.
- When PBS2 is ON, the counter preset is set to 5 using the MOVE instruction.
- When PBS3 is pressed as many times as the counter preset, Lamp1 lights up.
- When PBS4 is ON, the preset value is reset.

3. Address and Device Description

	Name	Function	I/O Allocation
Input	PBS1	Push Button Switch	X00
	PBS2	Push Button Switch	X01
	PBS3	Push Button Switch	X02
	PBS4	Push Button Switch	X03
Output	Lamp 1	Lamp	Y20

❒ PLC rendering

Practice Example 24 - Comparison operators

1. Practical Task

Understand the control circuit using comparison operators.

2. Operating Condition

- When PBS1 is ON, "100" is entered into D0.
- When PBS2 is ON, "200" is entered into D0.
- When PBS3 is ON, "300" is entered into D0.
- If the value of D0 is less than the reference value of 200, Lamp1 lights up.
- If the value of D0 is equal to the reference value of 200, Lamp2 lights up.
- If the value of D0 is greater than the reference value of 200, Lamp3 lights up.

3. Address and Device Description

	Name	Function	I/O Allocation
Input	PBS1	Push Button Switch	X00
	PBS2	Push Button Switch	X01
	PBS3	Push Button Switch	X02
Output	Lamp 1	Lamp	Y20
	Lamp 2	Lamp	Y21
	Lamp 3	Lamp	Y22

❐ PLC rendering

Practice Example 25 - INC operator

1. Practical Task

Understand the control circuit using the INC (Increment) operator.

2. Operating Condition

- When PBS1 (positive edge input) is pressed, D0 increases by "1".
- If the value equals "1" (using the [=] command), Lamp1 lights up.
- If the value equals "2" (using the [=] command), Lamp2 lights up.
- If the value equals "3" (using the [=] command), Lamp3 lights up.

3. Address and Device Description

	Name	Function	I/O Allocation
Input	PBS1	Push Button Switch	X00
	PBS2	Push Button Switch	X01
Output	Lamp 1	Lamp	Y20
	Lamp 2	Lamp	Y21
	Lamp 3	Lamp	Y22

❐ PLC rendering

Practice Example 26 - DEC operator

1. Practical Task

Understand the control circuit using the DEC (Decrement) operator.

2. Operating Condition

- When PBS1 is ON, "3" is entered into D0.
- When PBS2 (positive edge input) is pressed, the value decreases by "1".
- If the value equals "1" (using the [=] command), Lamp1 lights up.
- If the value equals "2" (using the [=] command), Lamp2 lights up.
- If the value equals "3" (using the [=] command), Lamp3 lights up.

3. Address and Device Description

	Name	Function	I/O Allocation
Input	PBS1	Push Button Switch	X00
	PBS2	Push Button Switch	X01
Output	Lamp 1	Lamp	Y20
	Lamp 2	Lamp	Y21
	Lamp 3	Lamp	Y22

❐ PLC rendering

Practice Example 27 - INC (Increment) + DEC (Decrement) application circuit

1. Practical Task

Control the lamps using the INC and DEC operators.

2. Operating Condition

- When PBS1 (positive edge input) is pressed, the value increases by "1".
- When PBS2 (positive edge input) is pressed, the value decreases by "1".
- If the value equals "3" (using the [=] command), Lamp1 lights up.

3. Address and Device Description

	Name	Function	I/O Allocation
Input	PBS1	Push Button Switch	X00
	PBS2	Push Button Switch	X01
Output	Lamp 1	Lamp	Y20

❐ PLC rendering

Practice Example 28 - INC (Increment) + DEC (Decrement) application circuit

1. Practical Task

Control the preset value of the counter using the INC and DEC operators.

2. Operating Condition

- When PBS1 (positive edge input) is pressed, the D0 value increases by "1".
- When PBS2 (positive edge input) is pressed, the D0 value decreases by "1".
- When PBS3 is toggled ON and OFF the number of times set in D0, Lamp1 lights up.
- When PBS4 is ON, the counter is reset.

3. Address and Device Description

	Name	Function	I/O Allocation
Input	PBS1	Push Button Switch	X00
	PBS2	Push Button Switch	X01
	PBS3	Push Button Switch	X02
	PBS4	Push Button Switch	X03
Output	Lamp 1	Lamp	Y20

❐ PLC rendering

Practice Example 29 - To control multiple lamps with a single switch.

1. Practical Task

To control multiple lamps with a single switch using a counter circuit.

2. Operating Condition

- Each time PBS1 is pressed, the lamps light up sequentially (the previous lamp turns off).
- When PBS2 is ON, all lamps turn off.

3. Address and Device Description

	Name	Function	I/O Allocation
Input	PBS1	Push Button Switch	X00
	PBS2	Push Button Switch	X01
Output	Lamp 1	Lamp	Y20
	Lamp 2	Lamp	Y21
	Lamp 3	Lamp	Y22
	Lamp 4	Lamp	Y23
	Lamp 5	Lamp	Y24
	Lamp 6	Lamp	Y25
	Lamp 7	Lamp	Y26
	Lamp 8	Lamp	Y27

❐ PLC rendering

Practice Example 30 - An emergency stop circuit using MC (Master Control) and MCR (Master Control Reset)

1. Practical Task

Understand the MC and MCR instructions.)

2. Operating Condition

- When PBS1 is ON, Lamp1 lights up, and even if PBS2 is ON, Lamp2 does not light up.
- When PBS1 is OFF, Lamp1 turns off, and if PBS2 is ON at that moment, Lamp2 lights up.

3. Address and Device Description

	Name	Function	I/O Allocation
Input	PBS1	Push Button Switch	X00
	PBS2	Push Button Switch	X01
Output	Lamp 1	Lamp	Y20
	Lamp 2	Lamp	Y21

❐ PLC rendering

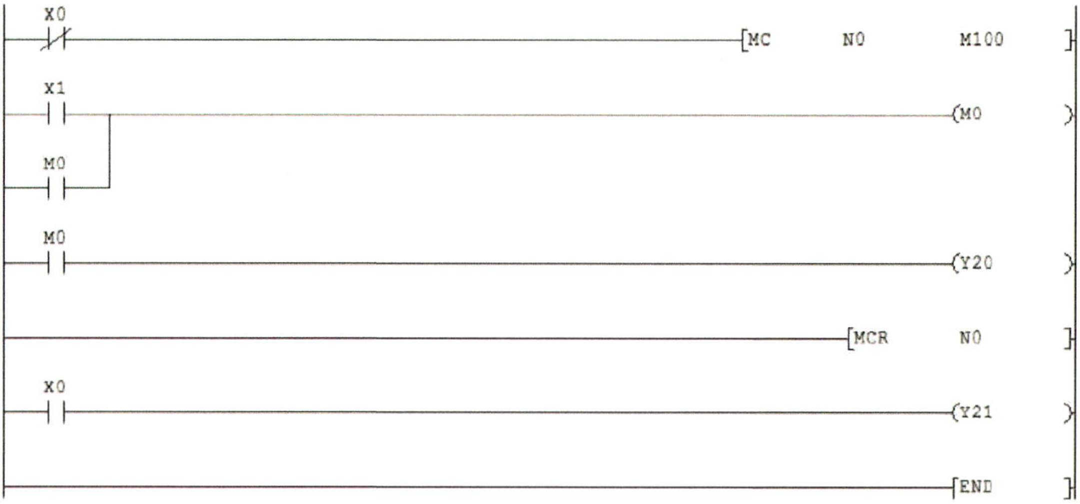

Practice Example 31 - Shift circuit

1. Practical Task

Control lamps using a shift circuit.

2. Operating Condition

- When PBS1 is ON, the lamps blink sequentially from Lamp1 to Lamp8 at 1-second intervals, continuously repeating the sequence from Lamp1 to Lamp8.
- When PBS2 is ON, the system resets.

3. Address and Device Description

	Name	Function	I/O Allocation
Input	PBS1	Push Button Switch	X00
	PBS2	Push Button Switch	X01
Output	Lamp 1 - 8	Lamp	Y20 - Y27

❐ PLC rendering

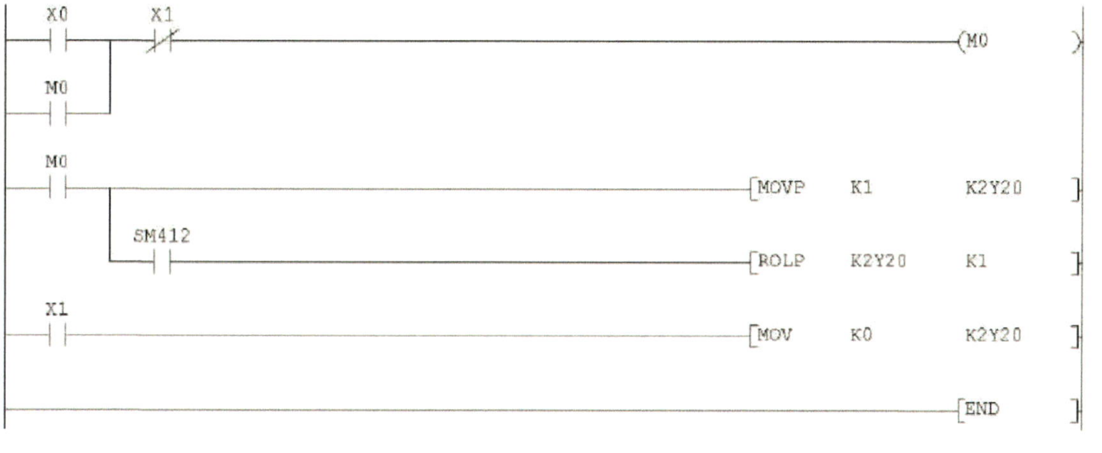

※ SM Instruction

Classification	Function Descriptions
SM400	Always ON device
SM401	Always OFF device
SM402	ON for 1 scan time after PLC starts, then OFF
SM403	OFF for 1 scan time after PLC starts, then ON
SM409	ON and OFF every 0.005 seconds
SM410	ON and OFF every 0.05 seconds
SM411	ON and OFF every 0.1 seconds
SM412	ON and OFF every 0.5 seconds

PLC Communication Settings

1. After completing the rendering and compiling (F4), select Online from the menu, and then click Specify Connection Target. A window will appear, as shown below, where you can configure the communication settings with the PLC.

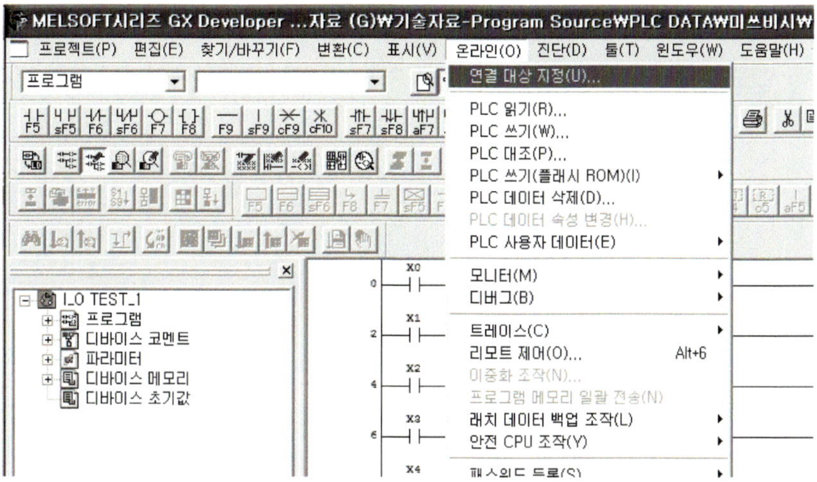

2. Click the Serial/USB button under PC I/F, and you will be able to select either the USB or Serial port as shown below. Choose the USB port.

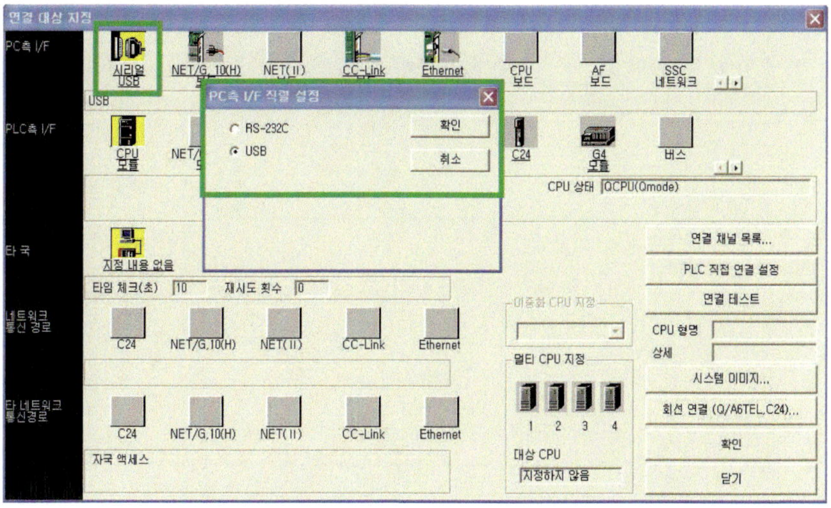

3. Perform a connection test to verify that the connection with the Q02(h) CPU is properly established. Once confirmed, click the OK button.

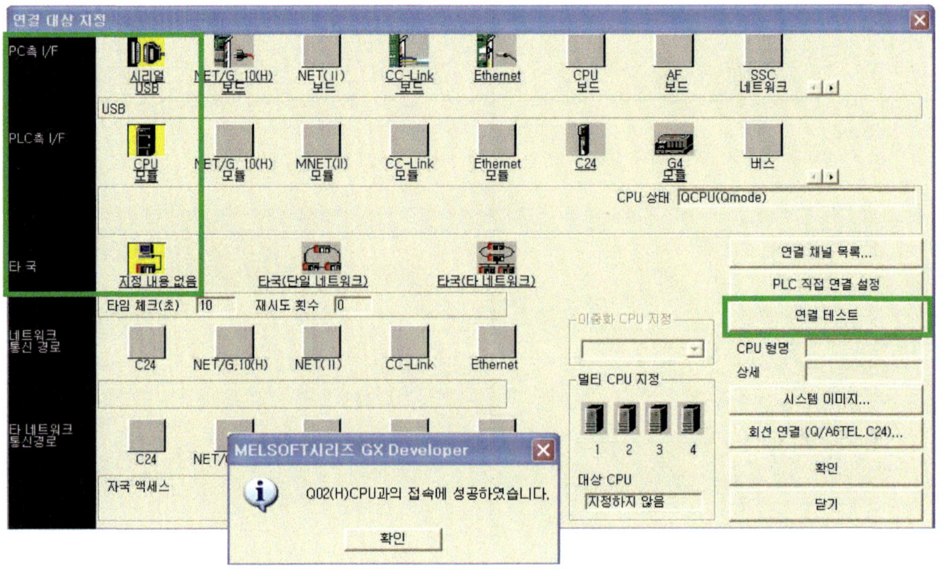

4. After a successful connection, select Write from the Online menu. Then, as shown in the diagram below, download the program to the PLC.

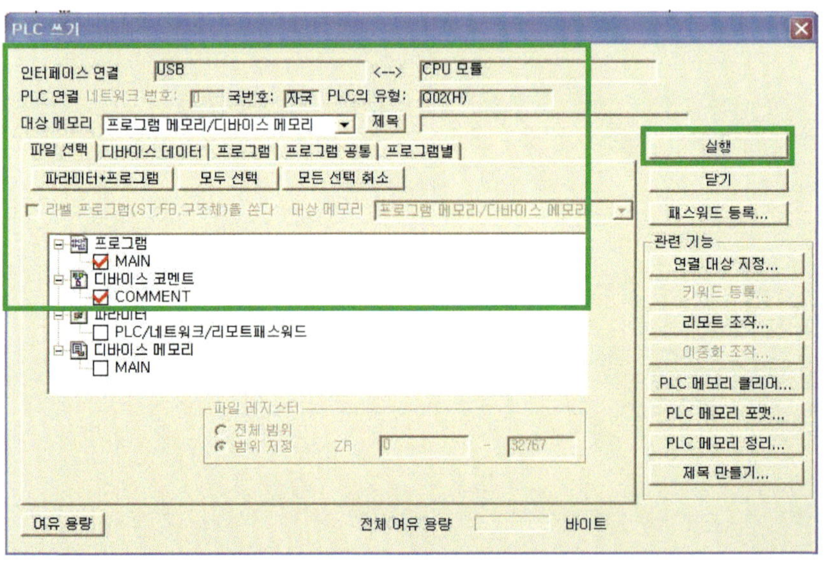

5. After selecting the items to download, click the Execute button to download the written program to the PLC.

6. Once the download is complete, press the shortcut key F2 or select Monitor -> Monitor Mode from the Online menu to observe the changes in the I/O.

3.
PLC trainer Practice Example

1 Practice: MITSUBISHI PLC Self-holding Circuit Practice

1. Operating Conditions

1) When X000 is activated, the output coil operates and becomes self-holding. At this time, the operation lamp lights up, and the stop lamp turns off.

2) When X001 is activated, the output coil stops operating. The program and circuit should be configured accordingly. At this time, the operation lamp turns off, and the stop lamp lights up.

3) The configuration of the PLC I/O MAP is as follows.

Category	Comment	I/O Allocation	Remarks
Input	Start Switch	X000	START
	Stop Switch	X001	STOP
Output	Output Coil	Y020	MOTOR
	Operation Lamp	Y021	Lamp L-1
	Stop Lamp	Y022	Lamp L-2

4) The input power is configured as DC 24V. After connecting the TRAINER COM to DC 24V(+), the switch COM is connected to DC 0V(-). The motor's power and the lamp are also configured using DC 24V, with DC 24V(+) connected to the TRAINER's output COM terminal and DC 0V(-) connected to the motor and lamp COM.

5) Configuring the I/O MAP is extremely important. The PLC I/O MAP is responsible for calculating the input and output points, as well as defining the input/output configuration and control.

6) Example program (Self-holding and motor driving circuit).

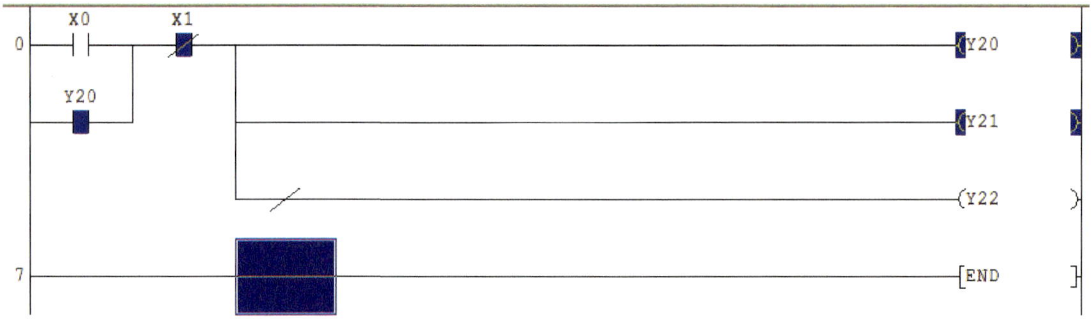

2. PLC Trainer Wiring Practice

1) After the program download is completed, perform the wiring as shown in the diagram below to conduct the I/O control practice.

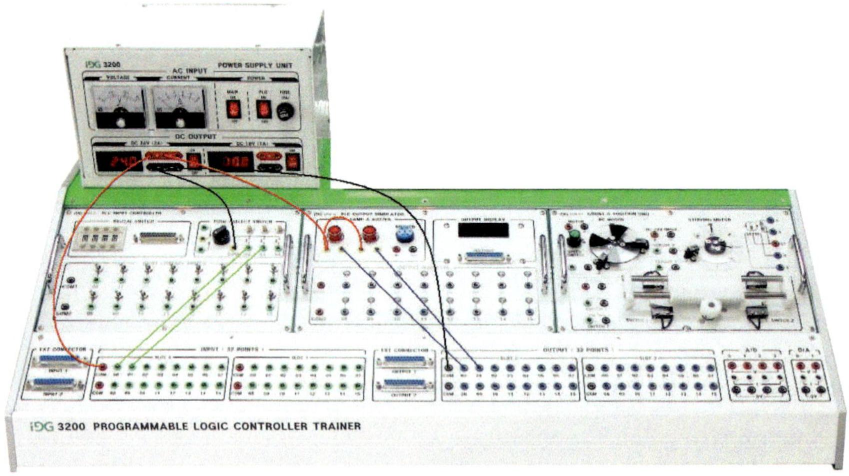

2) After wiring the DC 24V+ terminal to the PLC input COM terminal, connect the switch COM on top of the INPUT PUSH S.W. to 0V.
3) Once the connections are complete, toggle the PUSH switch On/Off to observe changes in the self-holding circuit.
4) Connect 0V to the PLC output COM terminal, then connect DC 24V to each lamp. Wire the remaining terminals of the lamps to the PLC output terminals.
5) You can monitor the changes in the I/O on the program screen of GX Developer.

2 Practice

MITSUBISHI PLC PLC Start-Stop Circuit Practice

1. Operating Conditions

1) Use a single input switch (start/stop) to configure a motor start circuit.(The outputs consist of an output coil for the contactor, an operation lamp, and a stop lamp).
2) When the input switch is toggled ON-OFF once, the motor starts, and the operation lamp lights up.
3) While running, if the input switch is toggled ON-OFF once again, the motor stops, the stop lamp lights up, and the operation lamp turns off.
4) The configuration of the I/O MAP is as follows.

Category	Comment	I/O Allocation	Remarks
Input	Start Switch	X0	START/STOP
Output	Motor(MC)	Y20	MOTOR
	Operation Lamp	Y21	Lamp L-1
	Stop Lamp	Y22	Lamp L-2

5) Example Program (Toggle Operation Circuit and Motor Driving Circuit)

2. PLC Trainer Wiring Practice

1) After the program download is completed, perform the wiring as shown in the diagram below to conduct the I/O control practice.

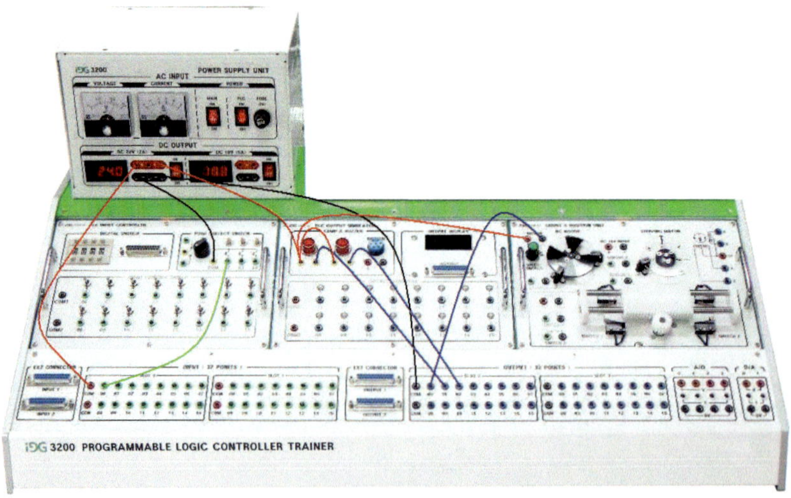

4

P r a c t i c e

Pulse Generation Program Using a Timer

1. Operating Conditions

1) When the input switch is pressed, Lamp 1 lights up for 1 second.
2) After 1 second, Lamp 1 turns OFF, activating Timer T2, which turns Lamp 1 back ON after a delay of 0.5 seconds.
3) When the stop switch is pressed, the lamp turns OFF immediately.
4) The configuration of the I/O MAP is as follows.

Category	Comment	I/O Allocation	Remarks
Input	Start Switch	X00	START
	Stop Switch	X01	STOP
Output	Lamp 1	Y20	Lamp 1

5) Example Program (Pulse Generation Circuit Using a Timer)

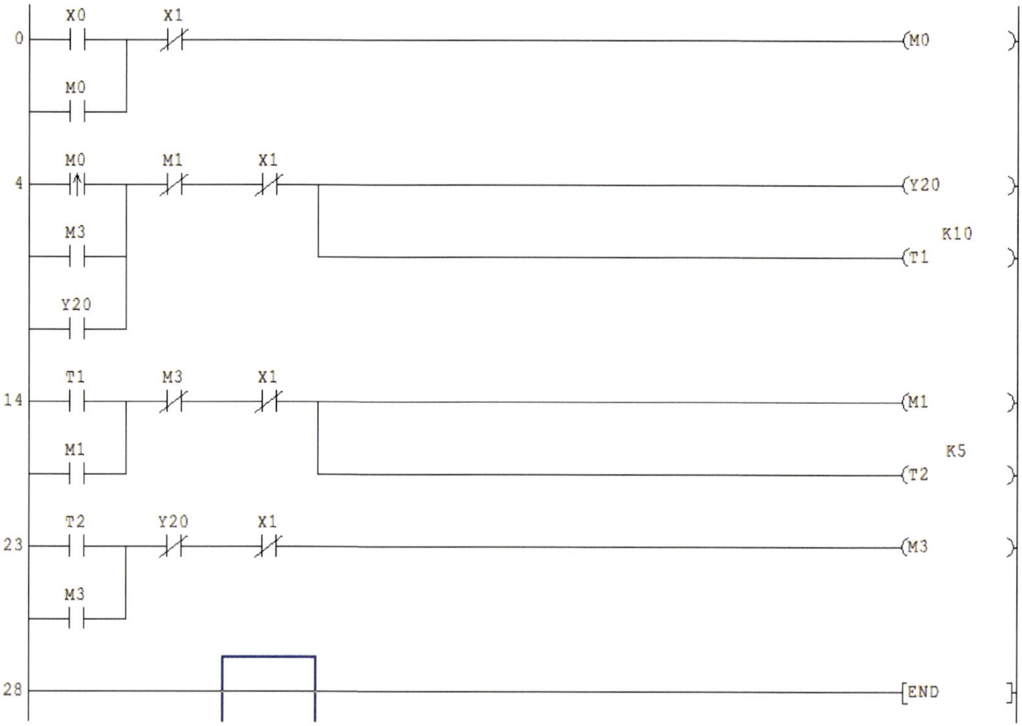

2. PLC Trainer Wiring Practice

1) After the program download is completed, perform the wiring as shown in the diagram below to conduct the I/O control practice.

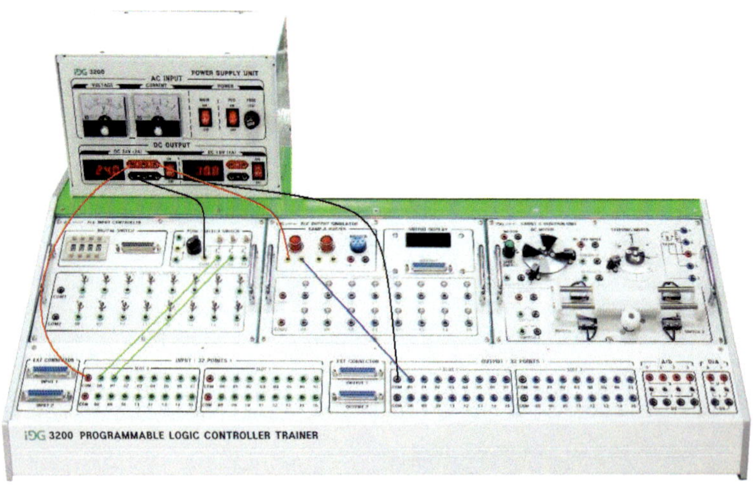

5 Practice
Lamp Control Program Using an UP Counter

1. Operating Conditions

1) When the input switch is toggled ON-OFF three times, Lamp 1 lights up. When toggled four times, Lamp 1 turns off.

2) The configuration of the I/O MAP is as follows.

Category	Comment	I/O Allocation	Remarks
Input	Start Switch	X00	
Output	Lamp 1	Y20	Lamp 1

3) Example Program (Lamp Control Circuit Using a Counter)

2. PLC Trainer Wiring Practice

1) After the program download is completed, perform the wiring as shown in the diagram below to conduct the I/O control practice.

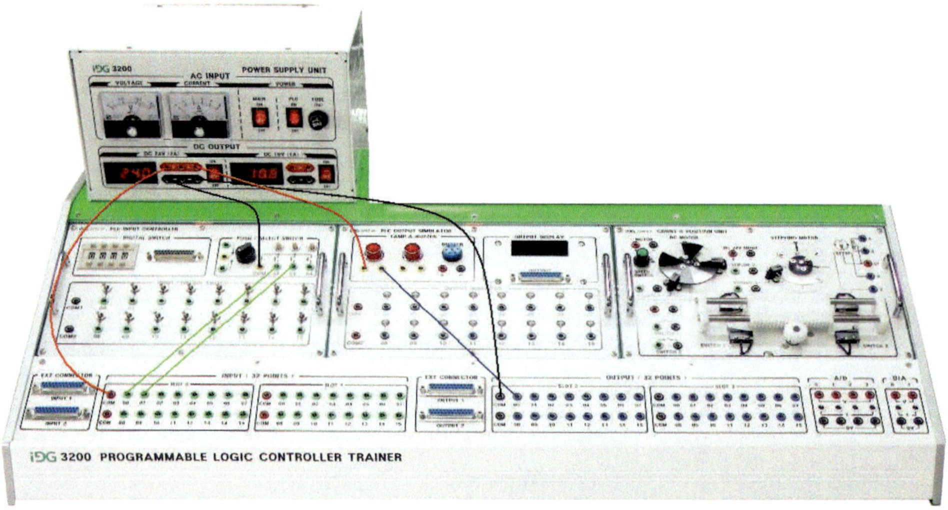

6 Practice
DC Motor Forward and Reverse Control Program

1. Operating Conditions

1) When the input switch (X00) is pressed, the DC motor rotates in the forward direction.

2) While the motor is rotating forward, the forward rotation lamp lights up.

3) When the stop switch is pressed, the motor stops, and the lamp turns off.

4) When the input switch (X01) is pressed, the DC motor rotates in the reverse direction.

5) While the motor is rotating in reverse, the reverse rotation lamp lights up.

6) The configuration of the I/O MAP is as follows.

Category	Comment	I/O Allocation	Remarks
Input	Forward Rotation Input Switch	X00	
	Reverse Rotation Input Switch	X01	
	Stop Switch	X02	
Output	Motor Forward Rotation A	Y20	Forward A
	Motor Forward Rotation B	Y30	Forward B
	Forward Rotation Lamp	Y22	Forward Lamp
	Motor Reverse Rotation A	Y31	Reverse A
	Motor Reverse Rotation B	Y21	Reverse B
	Reverse Lamp	Y23	Reverse Lamp

7) Example Program (DC Motor Forward and Reverse Control Circuit)

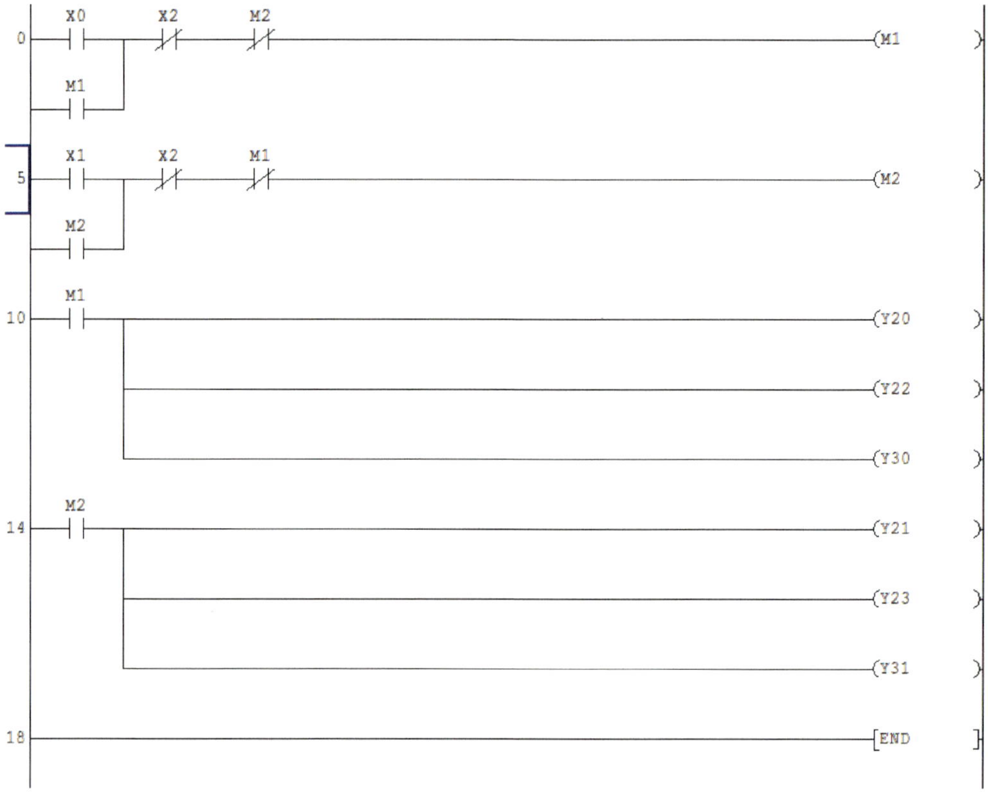

2. PLC Trainer Wiring Practice

1) After the program download is completed, perform the wiring as shown in the diagram below to conduct the I/O control practice.

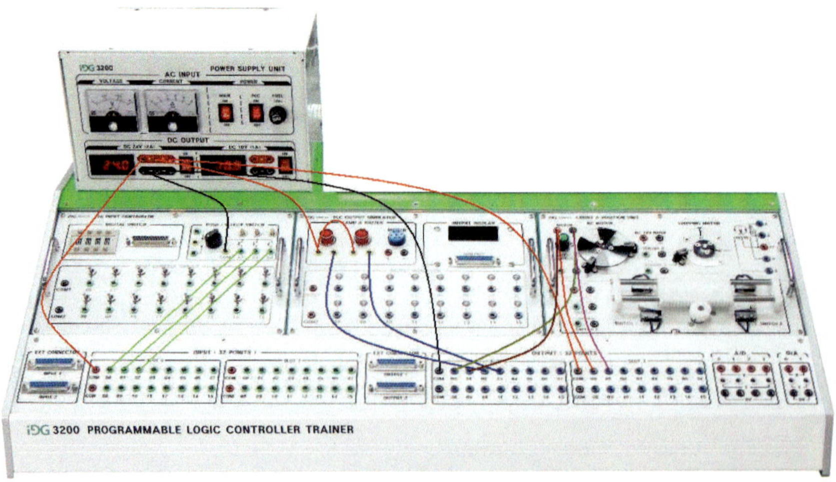

7 Practice
Motor Linear Motion Control Program (Forward/Reverse Drive Circuit)

1. Operating Conditions

1) When the input switch (X00) is pressed, the DC motor rotates in the forward direction.
2) If the motor detects a limit switch while rotating forward, it will rotate in the reverse direction.
3) If the motor detects a limit switch while rotating in reverse, it will rotate back in the forward direction.
4) When the input switch (X01) is pressed, the DC motor stops.
5) The configuration of the I/O MAP is as follows.

Category	Comment	I/O Allocation	Remarks
Input	Start Switch	X00	
	Stop Switch	X01	
	Right Limit Switch	X02	
	Left Limit Switch	X03	
Output	Forward Rotation Motor A	Y20	
	Forward Rotation Motor B	Y30	
	Reverse Rotation Motor A	Y31	
	Reverse Rotation Motor B	Y21	

6) Example Program (DC Motor Forward/Reverse Control Circuit)

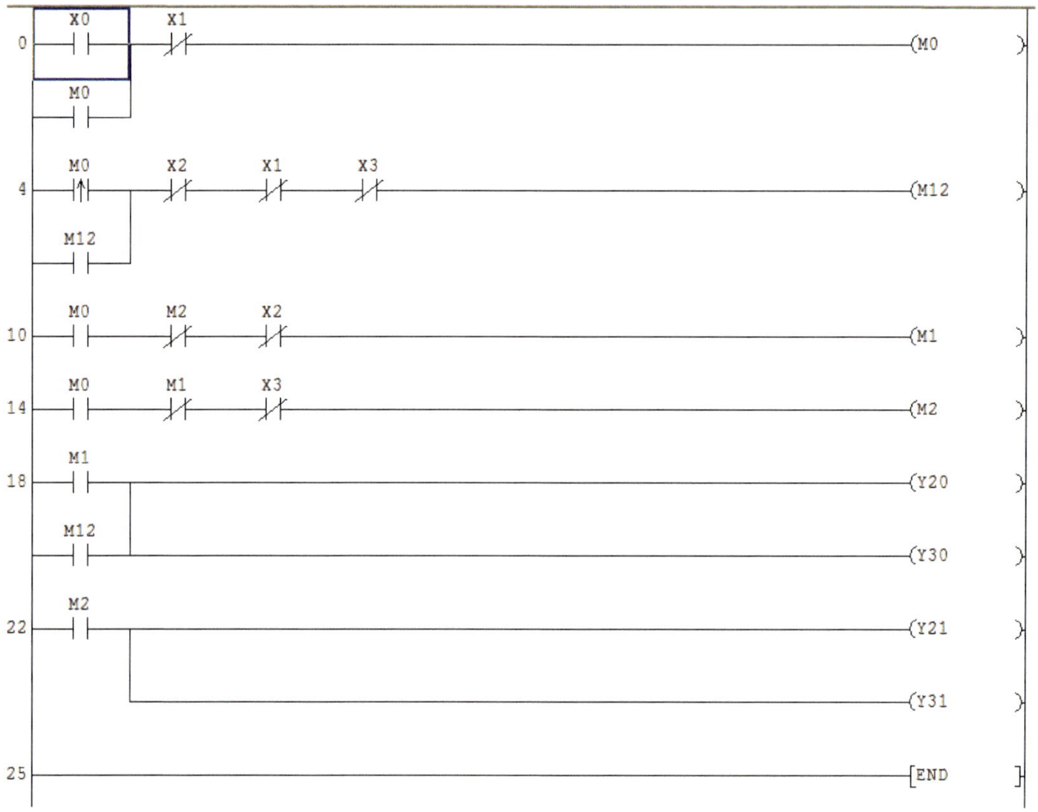

2. PLC Trainer Wiring Practice

1) After the program download is completed, perform the wiring as shown in the diagram below to conduct the I/O control practice.

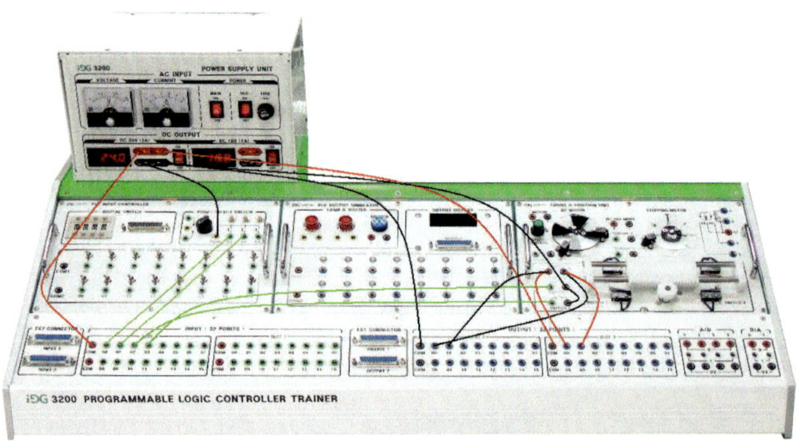

4. PLC Rendering Example

1 R e n d e r o n g
8-pin Relay

1. Structure and Contact Configuration of an 8-Pin Relay

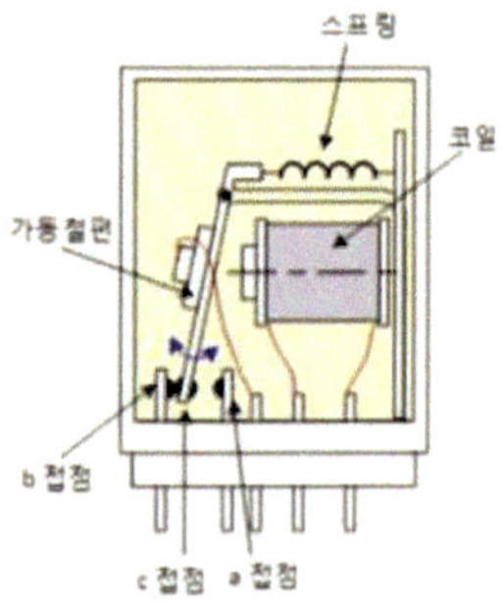

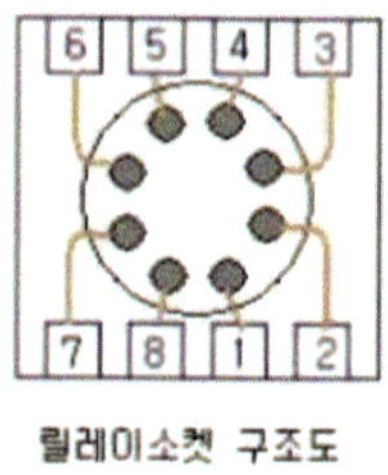

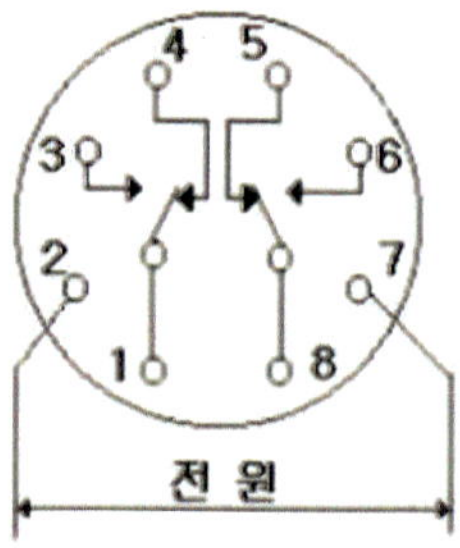

2. Sequence Diagram

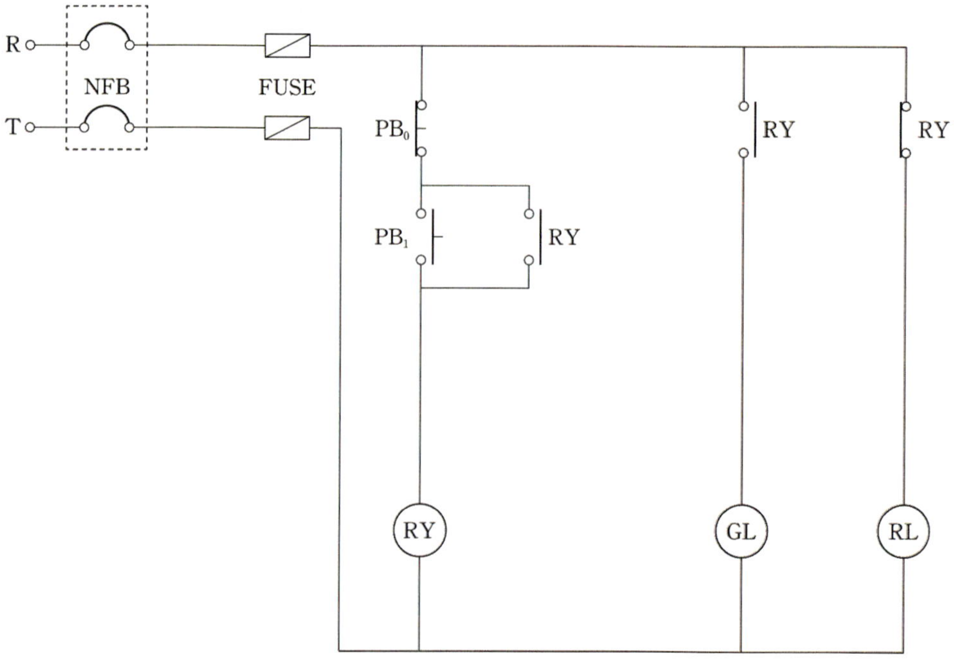

3. The I/O MAP

Input	I/O MAP	Output	I/O MAP
PB0 (Stop)	X00	RY (Relay power supply)	M0
PB1 (Start)	X01	GL (Start Lamp)	Y20
	X01	RL(Stop Lamp)	Y21

4. PLC Rendering

2 R e n d e r o n g
14-pin Relay

1. Structure and Contact Configuration of an 14-Pin Relay

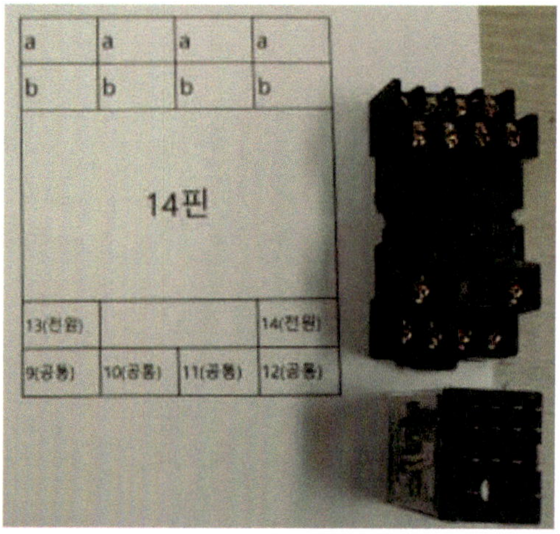

13-14 Power

9-5 a Contact, 9-1 b Contact

10-6 a Contact, 10-2 b Contact

11-7 a Contact, 11-3 b Contact

12-8 a Contact, 12-4 b Contact

2. Sequence Diagram

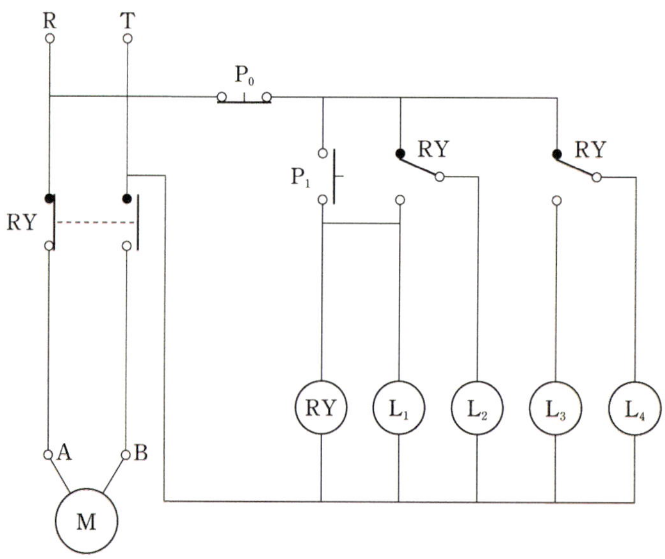

3. The I/O MAP

Input	I/O MAP	Output	I/O MAP
P0 (Stop)	X00	RY (Relay power supply)	M0
P1 (Start)	X01	L1	Y20
		L2	Y21
		L3	Y22
		L4	Y23

4. PLC Rendering

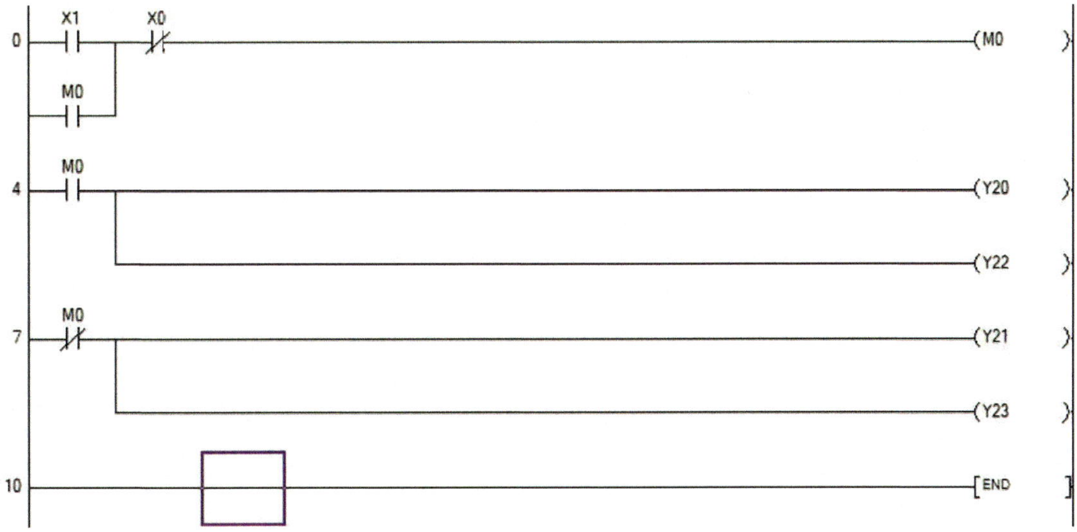

3 Rendering

Timer Sequence Program Conversion

1. Sequence Diagram

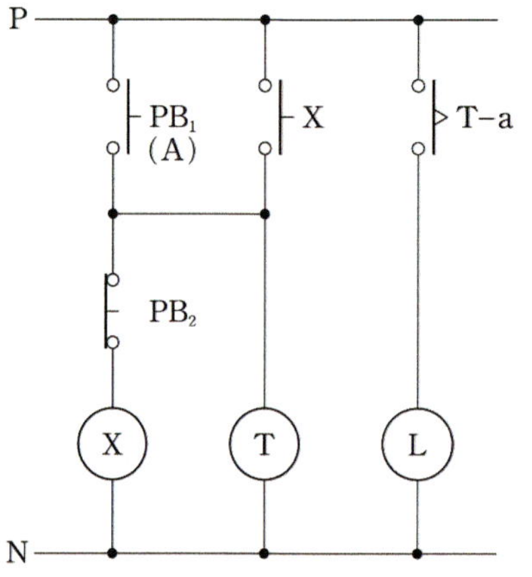

2. The I/O MAP

Input	I/O MAP	Output	I/O MAP
BS1 (Start)	X00	X(Relay power supply)	M0
BS2 (Stop)	X01	T (Timer Setting: 3 Sec)	T0
		L	Y20

3. PLC Rendering

4

R e n d e r o n g

Timer Sequence Program Conversion

1. Sequence Diagram

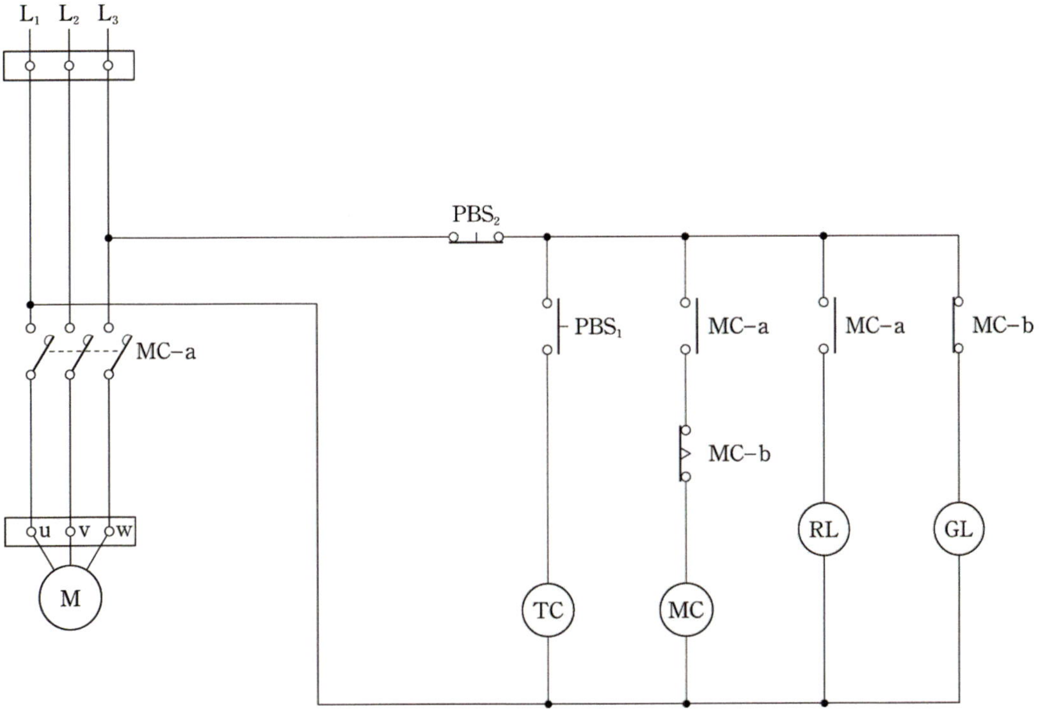

2. The I/O MAP

Input	I/O MAP	Output	I/O MAP
PBS1 (Start)	X00	MC (Internal Device Memory)	M0
PBS2 (Stop)	X01	TC (Timer Setting: 3 Sec)	T0
		RL	Y20
		GL	Y21

3. PLC Rendering

5 R e n d e r o n g
Timer Sequence Program Conversion

1. Sequence Diagram

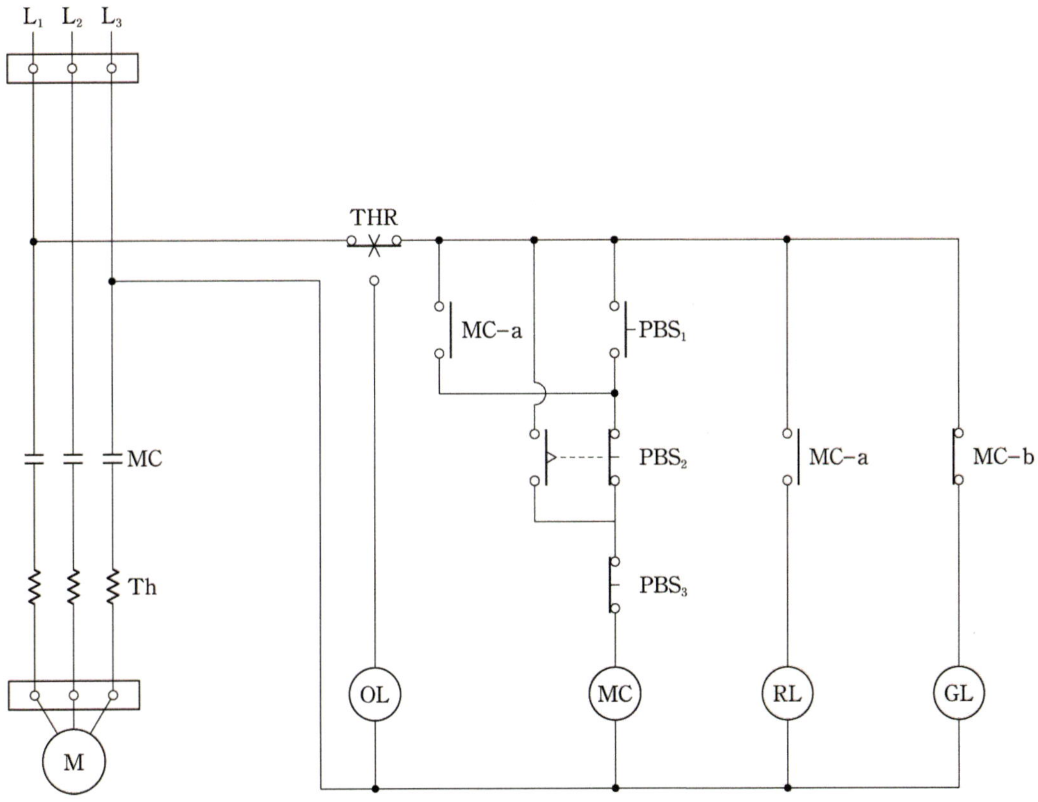

2. The I/O MAP

Input	I/O MAP	Output	I/O MAP
PBS1 (Start)	X00	MC (Internal Device Memory)	M0
PBS3 (Stop)	X01	RL	Y20
PBS2 (Chattering)	X02	GL	Y21

3. PLC Rendering

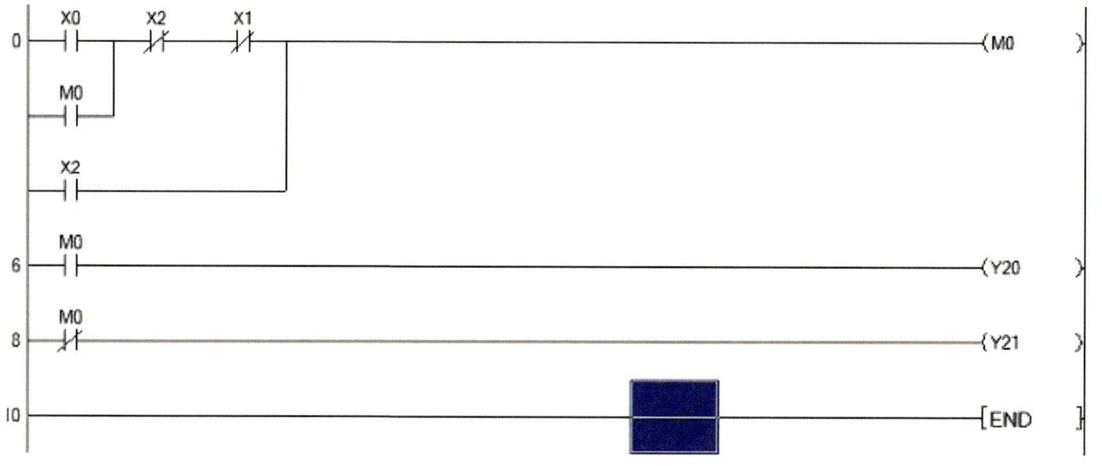

6 R e n d e r o n g
Motor Forward/Reverse Control Circuit

1. Sequence Diagram

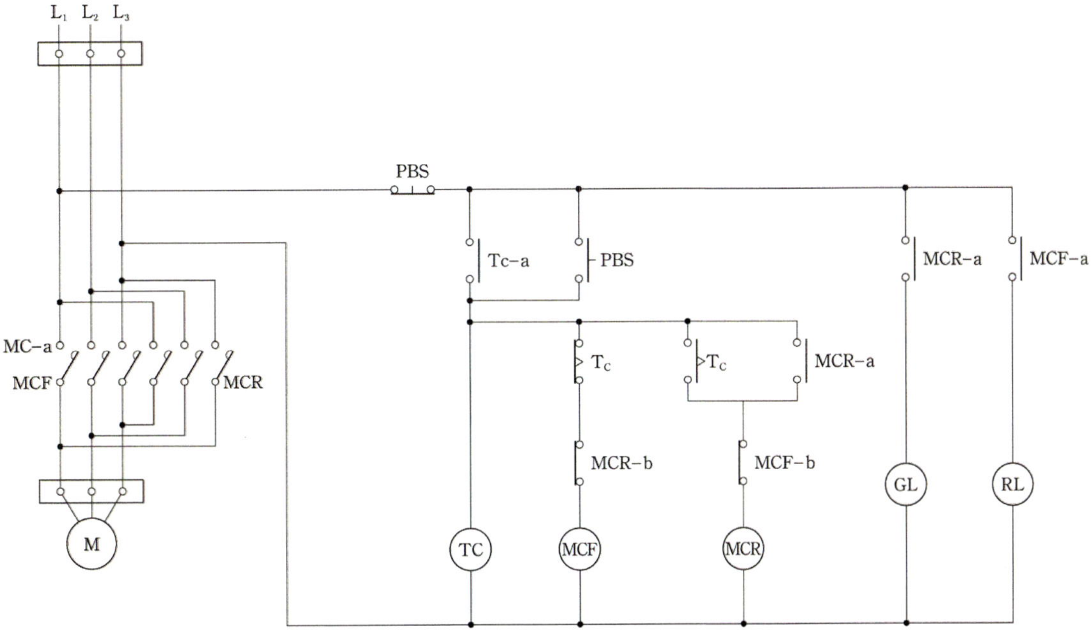

2. The I/O MAP

Input	I/O MAP	Output	I/O MAP
PBS0 (Stop)	X00	Virtual Device Memory	M2
PBS1 (Start)	X01	MCF (Forward)	M0
		MCR (Reverse)	M1
		TC (Timer Setting: 3 Sec)	T0
		GL (Reverse Lamp)	Y20
		RL (Forward Lamp)	Y21

3. PLC Rendering

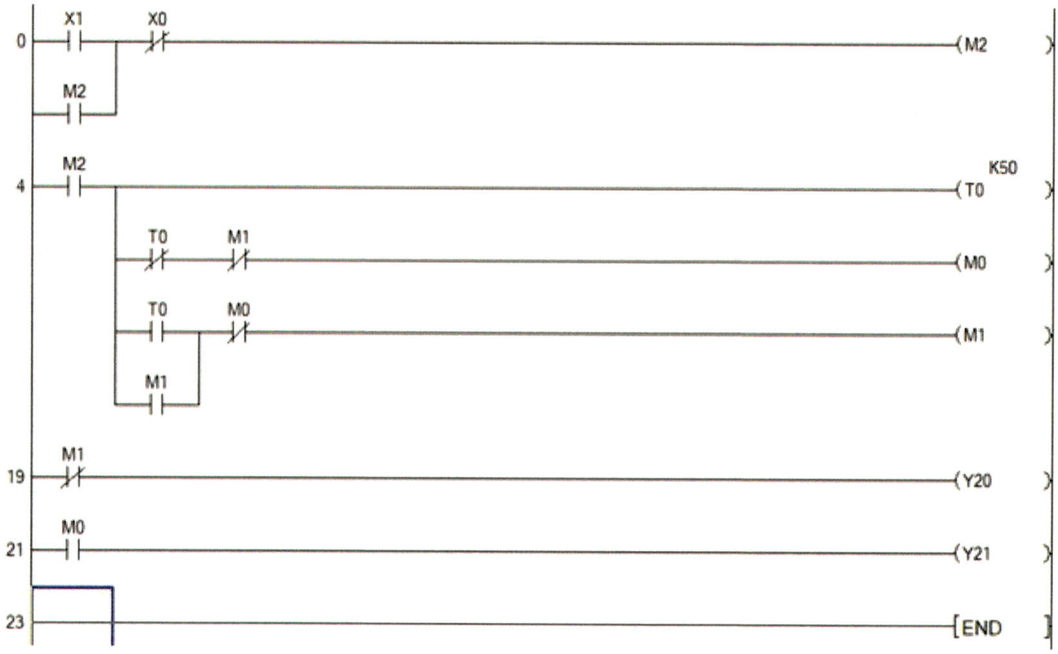

7 R e n d e r o n g

Motor Forward/Reverse Control Circuit 2

1. Sequence Diagram

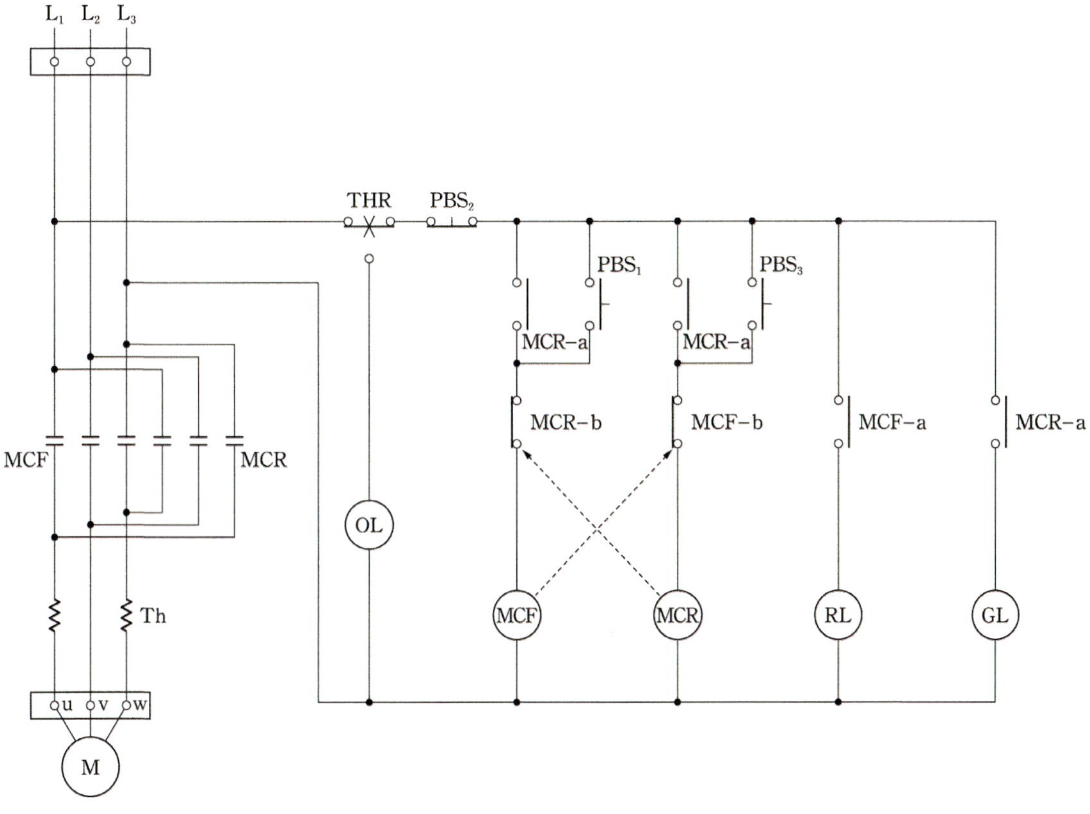

2. The I/O MAP

Input	I/O MAP	Output	I/O MAP
PBS1 (Forward)	X00	MCF (Forward)	Y20
PBS2 (Reverse)	X01	MCR (Reverse)	Y21
PBS0 (Stop)	X00	RL (Forward Lamp)	Y30
		GL (Reverse Lamp)	Y31

3. PLC Rendering

8 Rendering
Dependent Circuit

1. Sequence Diagram

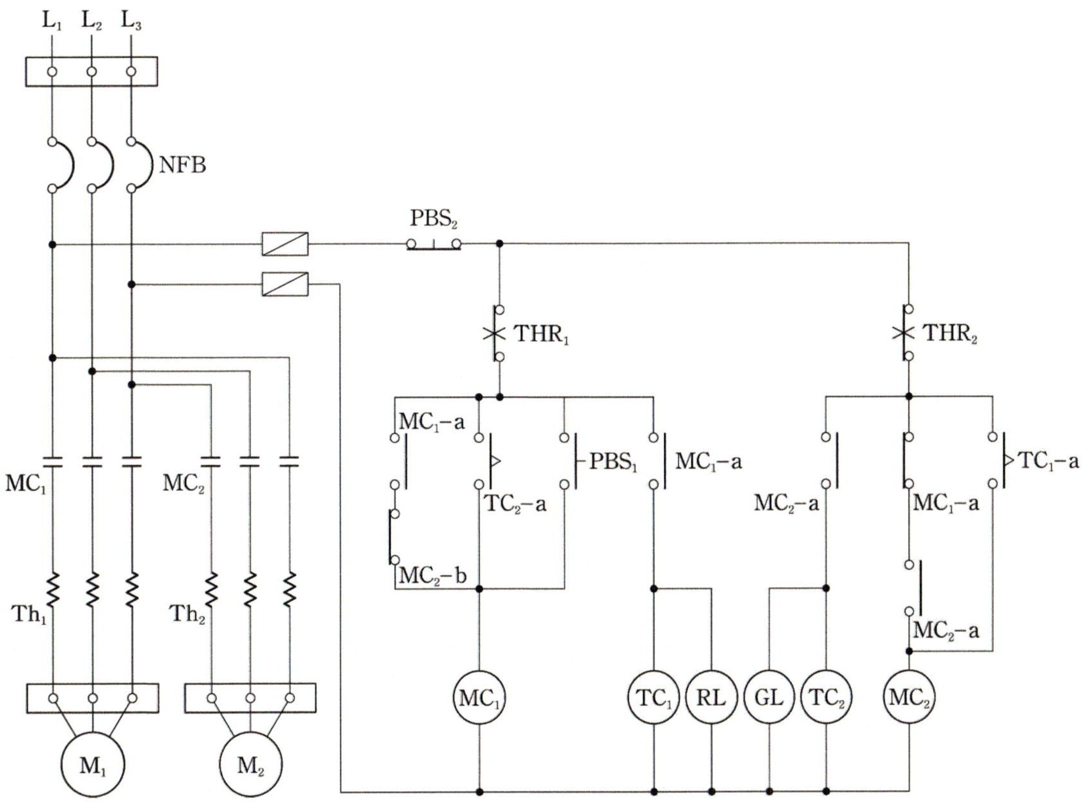

2. The I/O MAP

Input	I/O MAP	Output	I/O MAP
PBS1 (Start)	X00	MC1	M0
PBS2 (Stop)	X01	MC2	M1
		TC1 (Timer Setting: 3 Sec)	T0
		TC2 (Timer Setting: 3 Sec)	T1
		RL	Y20
		GL	Y21

3. PLC Rendering

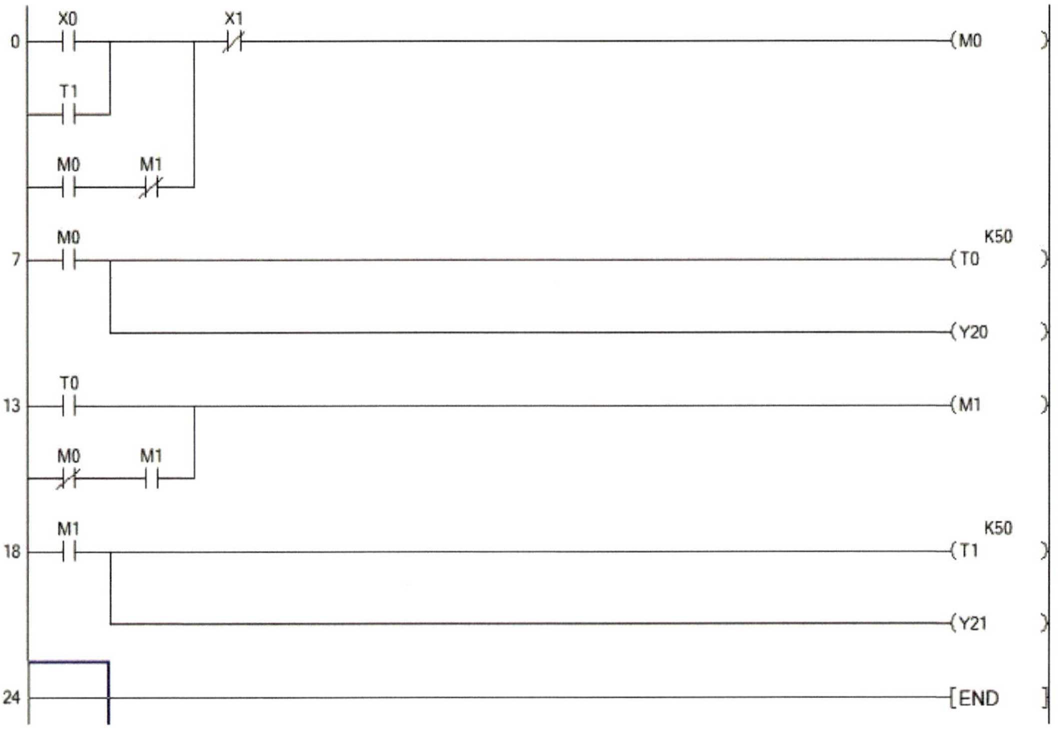

R e n d e r o n g

9 Automatic Forward/Reverse Motor Starting Circuit Using Three Timers

1. Sequence Diagram

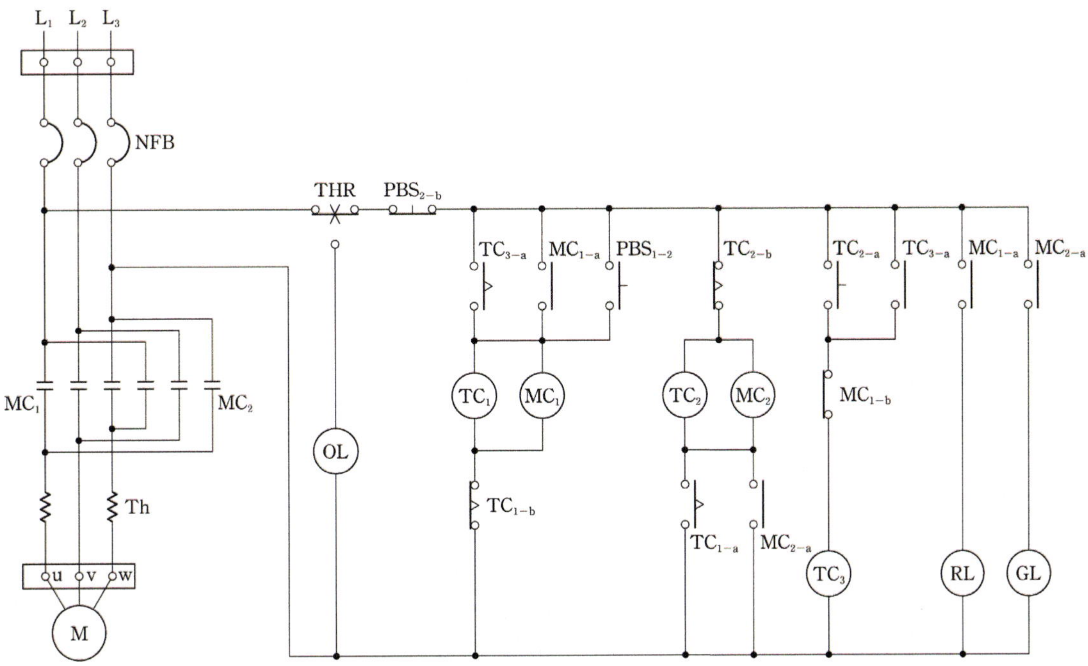

2. The I/O MAP

Input	I/O MAP	Output	I/O MAP
PBS1 (Start)	X00	Virtual Device Memory	M0
PBS2 (Stop)	X01	Virtual Device Memory	M1
		MC1	M2
		MC2	M3
		TC1 (Timer Setting: 3 Sec)	T0
		TC2 (Timer Setting: 3 Sec)	T1

		TC3 (Timer Setting: 3 Sec)	T2
		RL	Y20
		GL	Y21

3. PLC Rendering

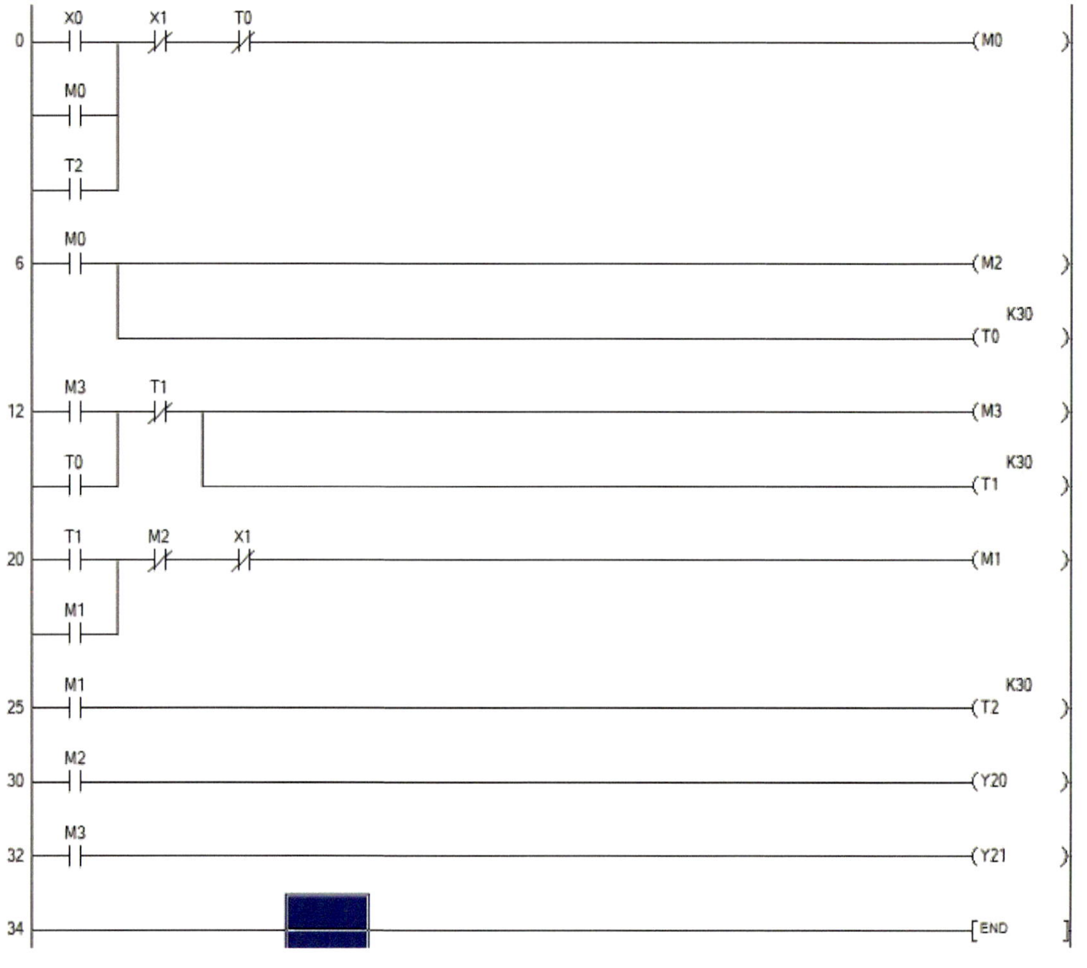

10 Rendering — Three-Way Switch Circuit

1. Sequence Diagram

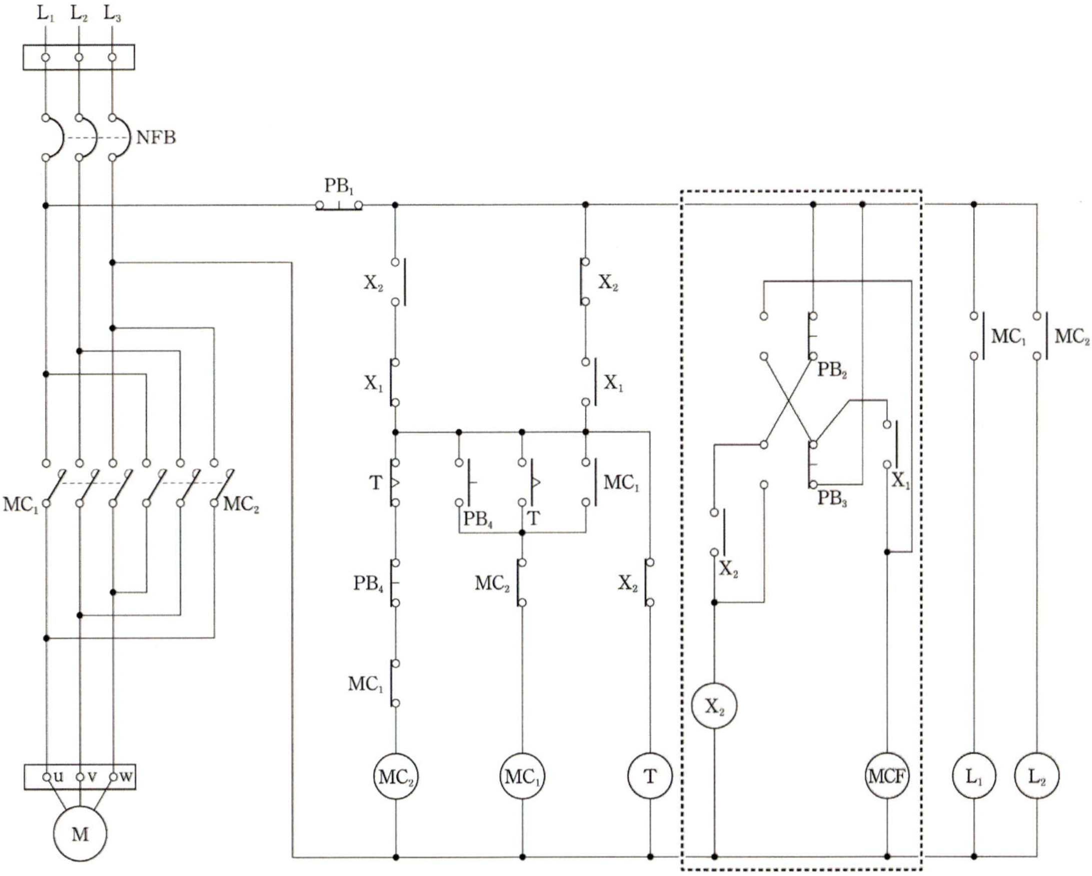

2. Three-Way Switch Section (Red) - Rendering Conversion

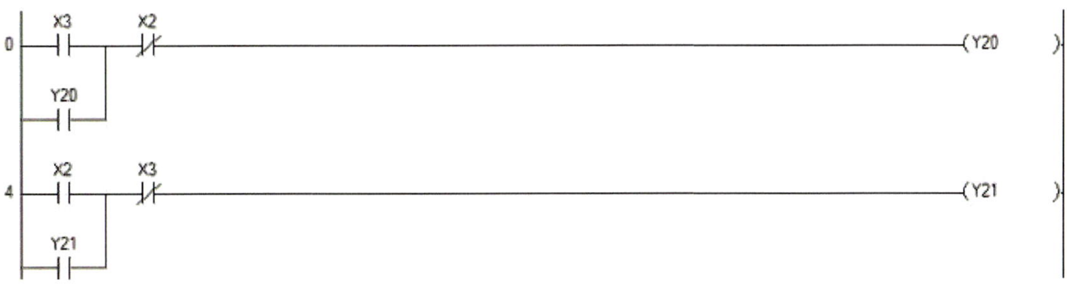

11 Rendering
Y-Δ Connection

1. Sequence Diagram

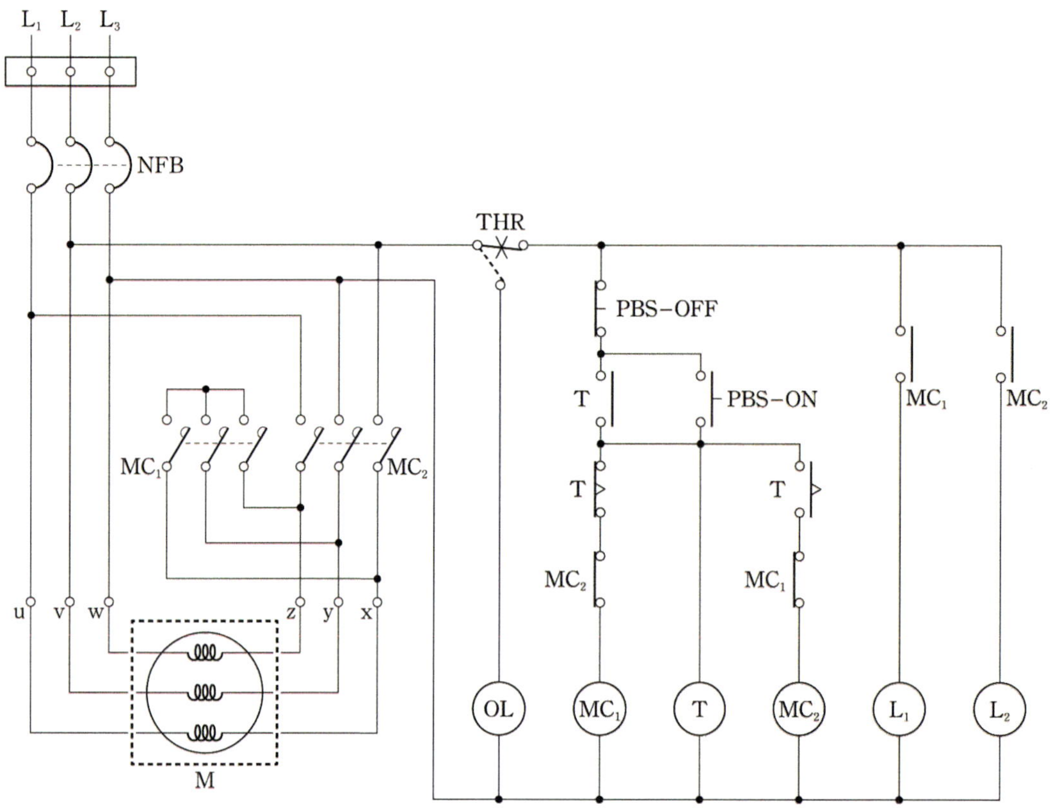

2. The I/O MAP

Input	I/O MAP	Output	I/O MAP
PBS-ON	X00	Virtual Device Memory	M0
PBS2-OFF	X01	MC1 (Δ)	M1
		MC2 (Y)	M2
		T (Timer Setting: 3 Sec)	T0
		L1 (Lamp)	Y20
		L2 (Lamp)	Y21

3. PLC Rendering

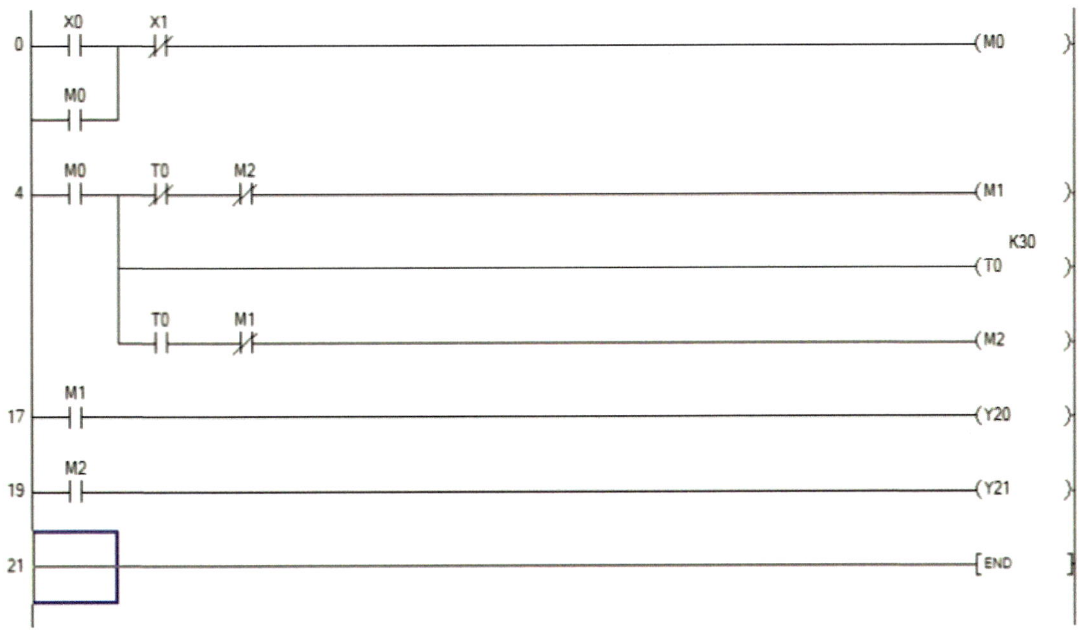

R e n d e r o n g

12 Y-Δ Starting Circuit Using Two Magnetic Contactors

1. Sequence Diagram

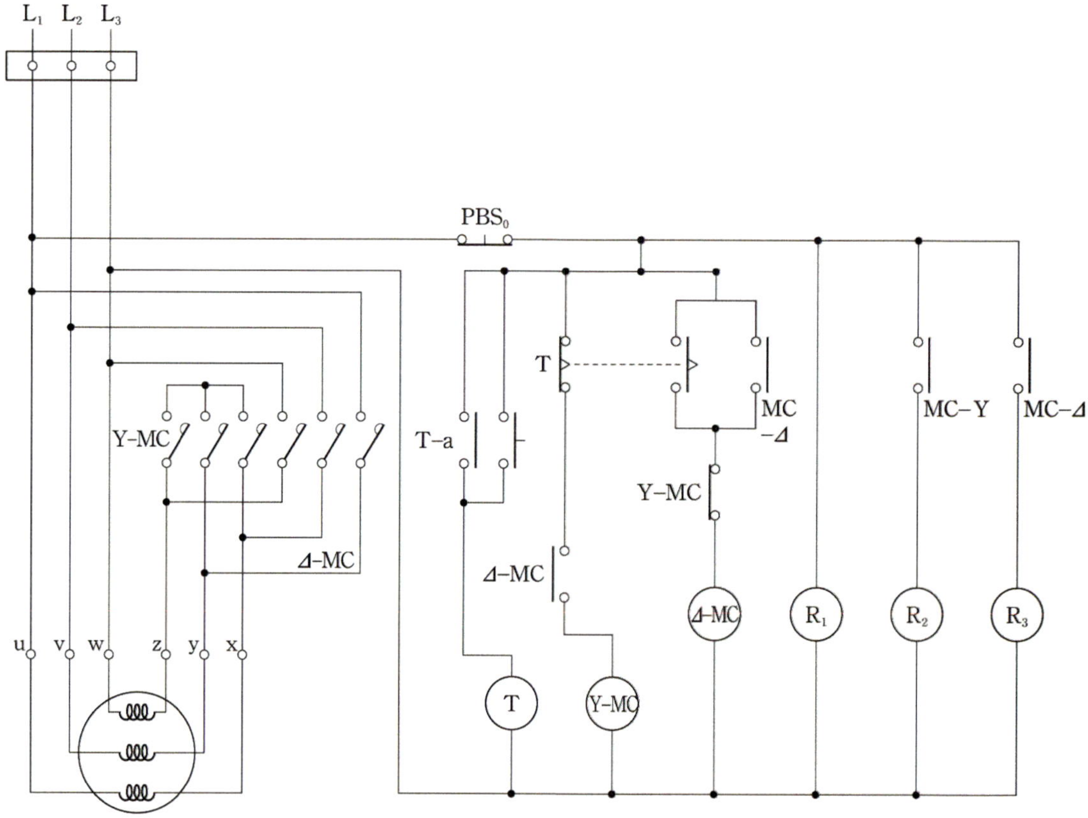

2. The I/O MAP

Input	I/O MAP	Output	I/O MAP
PBS1 (Start)	X00	Timer of Virtual Device Memory	M2
PBS0 (Stop)	X01	T (Timer Setting: 3 Sec)	T0
		Y- MC	M0
		Δ - MC	M1
		R1	Y20
		R2	Y21
		R3	Y22

3. PLC Rendering

13 R e n d e r o n g

Forward/Reverse Motor Control Using Limit Switches

1. Sequence Diagram

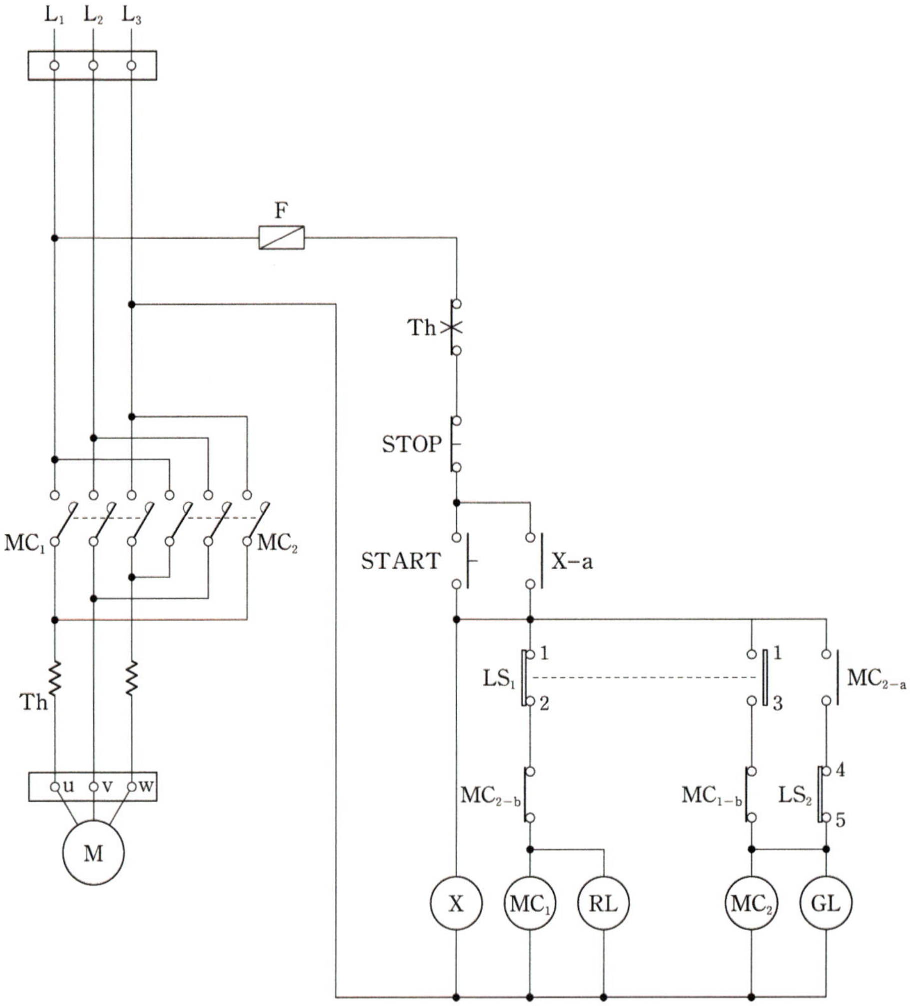

2. The I/O MAP

Input	I/O MAP	Output	I/O MAP
Start	X00	X	M0
Stop	X01	MC1 (Forward)	M1
LS1	X02	MC2 (Reverse)	M2
LS2	X03	RL	Y20
		GL	Y21

3. PLC Rendering

14 Rendering
PLC Ladder Diagram Using a Time Chart 1

1. Time Chart

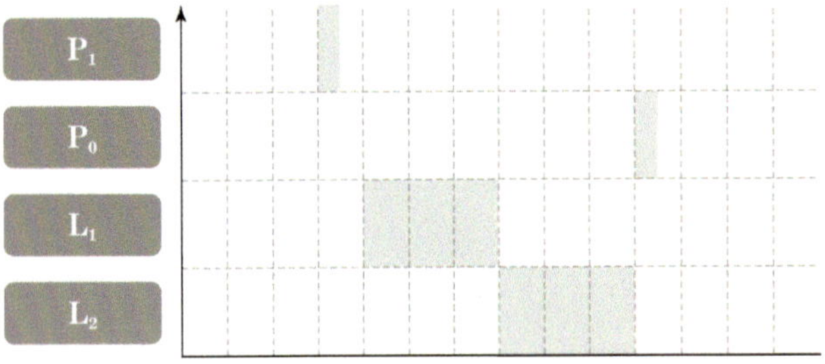

2. The I/O MAP

Input	I/O MAP	Output	I/O MAP
P1 (Start)	X00	L1	Y20
P0 (Stop)	X01	L2	Y21
		Timer (Timer Setting: 3 Sec)	T0

3. PLC Rendering

15 R e n d e r o n g
PLC Ladder Diagram Using a Time Chart 2

1. Time Chart

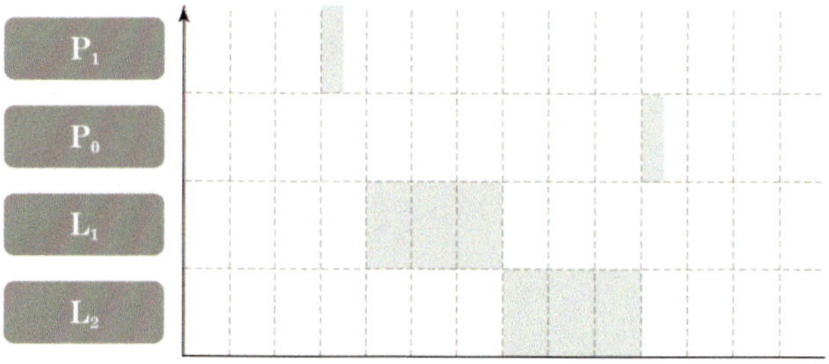

2. The I/O MAP

Input	I/O MAP	Output	I/O MAP
P1 (Start)	X00	L1	Y20
P0 (Stop)	X01	L2	Y21
		Timer (Timer Setting: 3 Sec)	T0

3. PLC Rendering

R e n d e r o n g

16 PLC Ladder Diagram Using a Time Chart 3

1. Time Chart

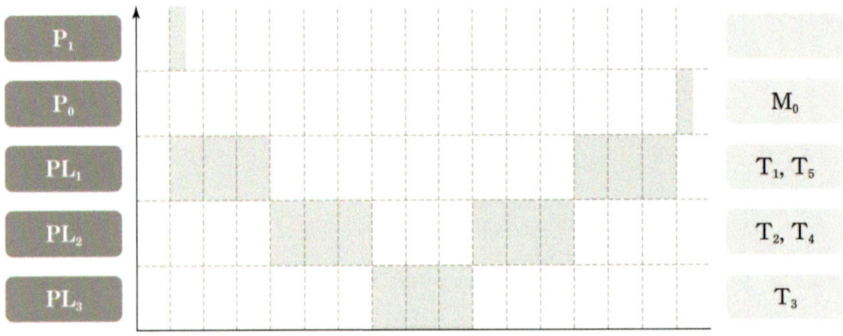

2. Operation Description

(1) When P1 (X00) is turned ON, the PL1, PL2, and PL3 lamps blink sequentially at 3-second intervals.

(2) When P0 (X01) is input, the system is reset to the initial state.

(3) During blinking, the sequence PL1 (Y20) → PL2 (Y21) → PL3 (Y22) → PL2 → PL1 is repeated three times.

3. PLC Rendering

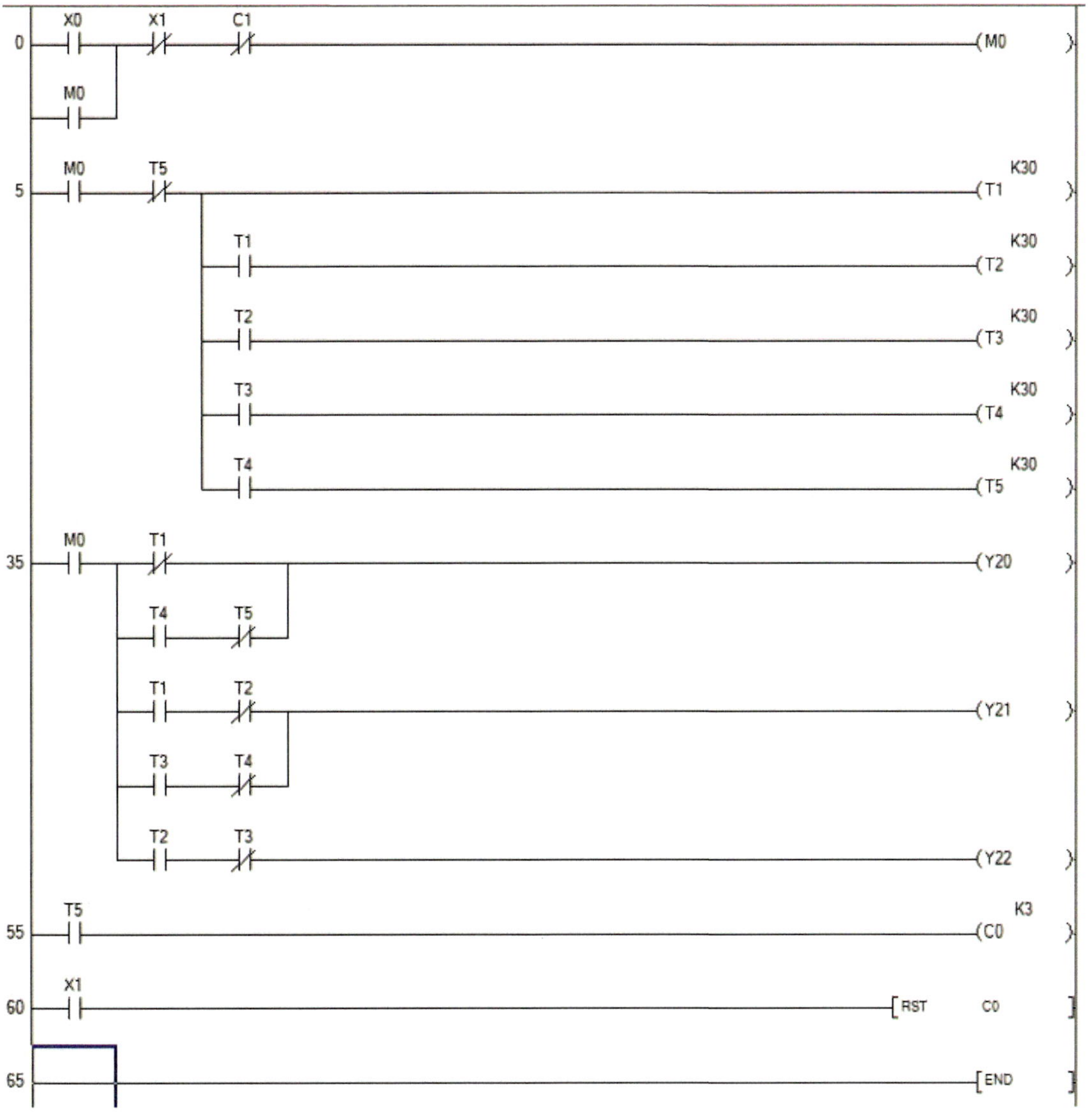

17

R e n d e r o n g

PLC Ladder Diagram Using a Time Chart 4

1. Time Chart

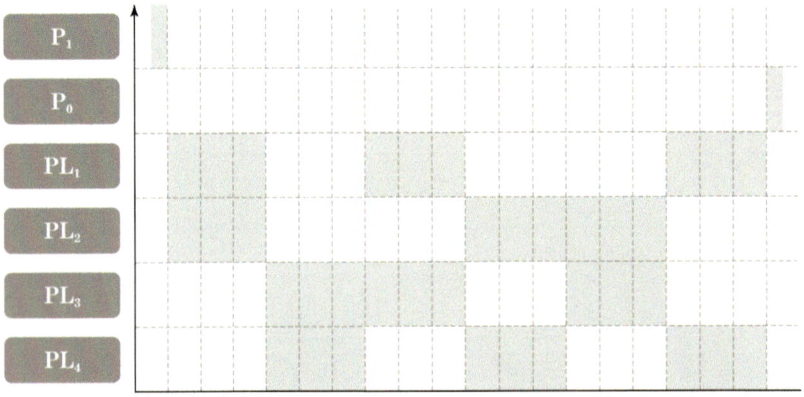

2. Operating Conditions

(1) When P1 is activated, PL1, PL2, PL3, and PL4 blink two lamps at a time sequentially at 3-second intervals.

(2) When P0 is activated, the system is reset to the initial state.

(3) When P1 is activated, the lamp sequence operates continuously with a 3-second cycle.

(4) When P1 is activated, two lamps turn ON simultaneously in the following order: PL1/PL2 → PL3/PL4 → PL1/PL3 → PL2/PL4 → PL2/PL3 → PL1/PL4, and this sequence operates at 3-second intervals, repeated three times.

3. The I/O MAP

Input	I/O MAP	Output	I/O MAP
P1 (Start)	X00	Virtual Device Memory	M0
P0 (Stop)	X01	Counter function	C0
		6 Timers (Timer Setting: 3 Sec)	T0~T5

		PL1	Y20
		PL2	Y21
		PL3	Y22
		PL4	Y23

4. PLC Rendering

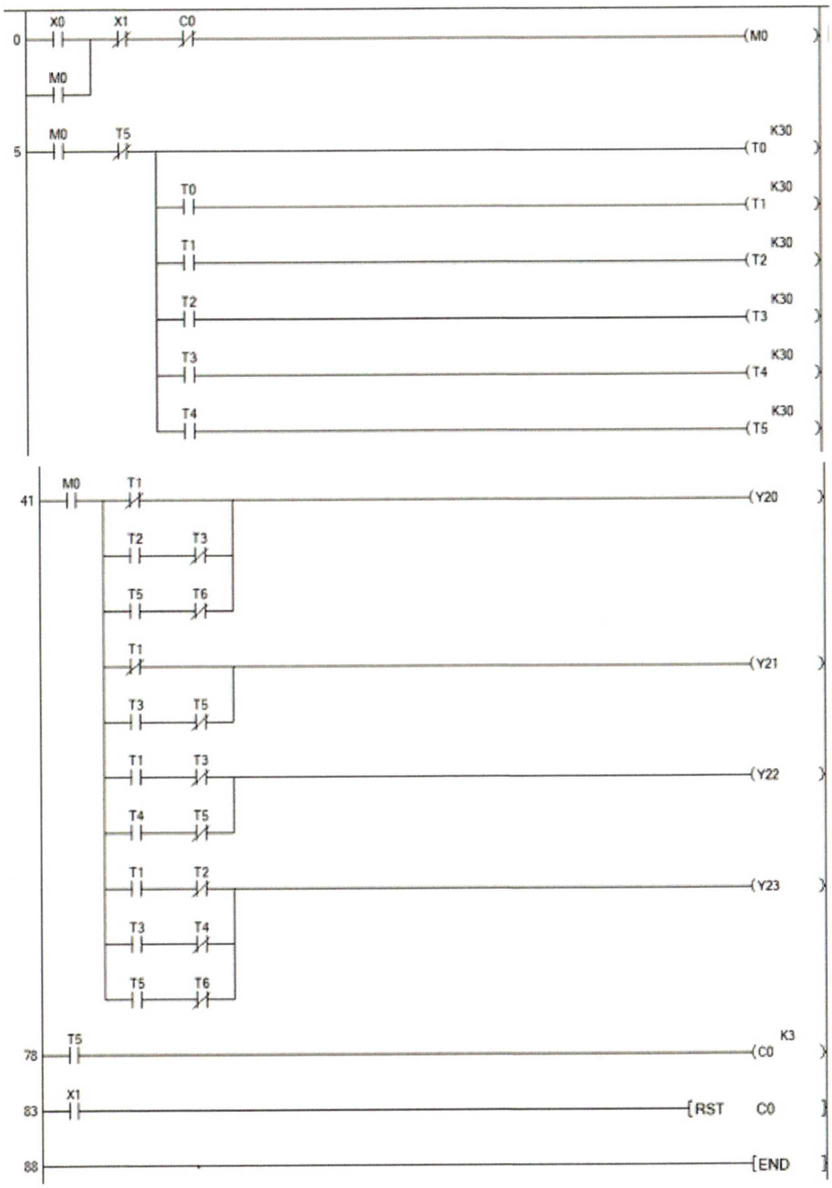

18 R e n d e r o n g Quiz Program

1. Operation Description

(1) When either Contestant 1 or 2 presses a switch, self-holding and mutual interlocking are activated.

(2) When the contestant presses the button, the buzzer sounds four times at 1-second intervals. (Using SM412 & Counter)

(3) After 5 seconds of the contestant's button operation, the system resets, and the contestant's lamp turns off to return to the initial state. (Using a timer)

(4) When the host presses the reset button, the contestant's lamp turns off, and the system is reset.

2. The I/O MAP

Input	I/O MAP	Output	I/O MAP
Contestants 1	X00	Contestants 1 Lamp	Y20
Contestants 2	X01	Contestants 2 Lamp	Y21
Host Reset	X02	buzzer	Y22

3. PLC Rendering

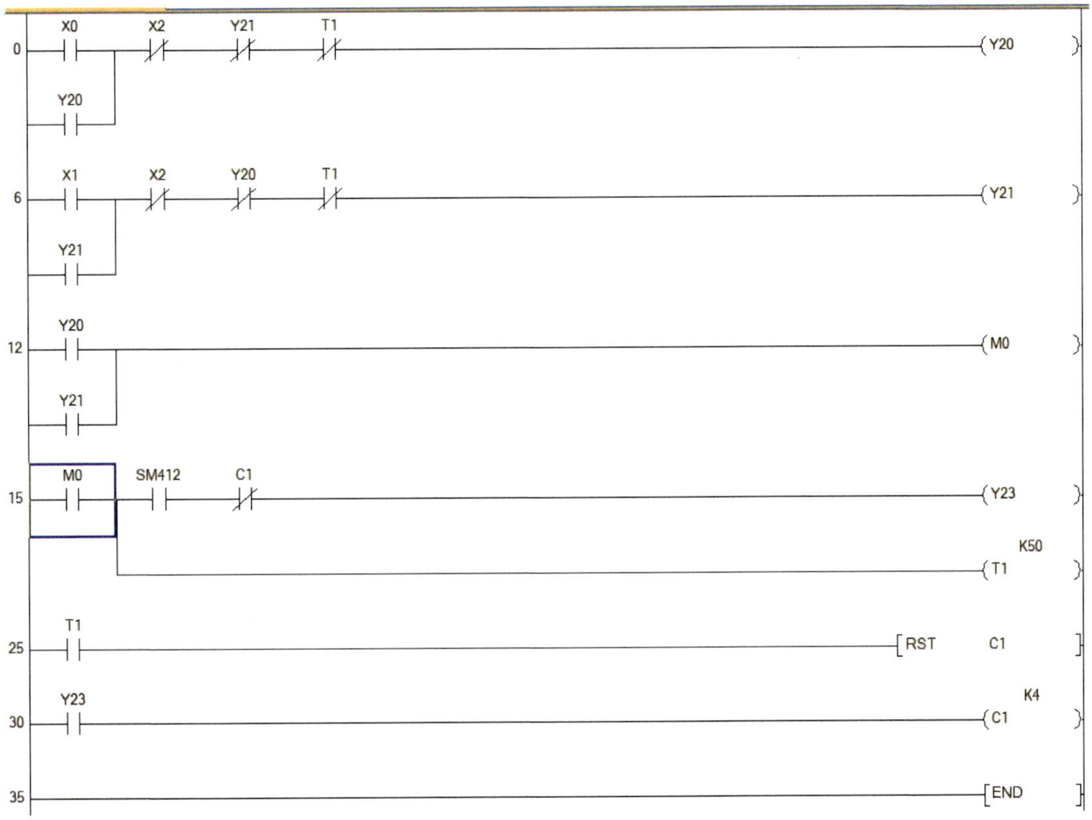

19 R e n d e r o n g
Motor condition monitoring

1. Operation Description

The process of rotating the motor, sensing 5 times, and repeating the cycle where the motor stops for 5 seconds

2. The I/O MAP

Input	I/O MAP	Output	I/O MAP
Start	X00	Self-holding Relay	m0
Stop	X01	Motor Running Lamp	Y21
Sensor -|↑|- sF7	X02 (Pulse Rising)	Motor Stopping Lamp	Y22

3. PLC Rendering

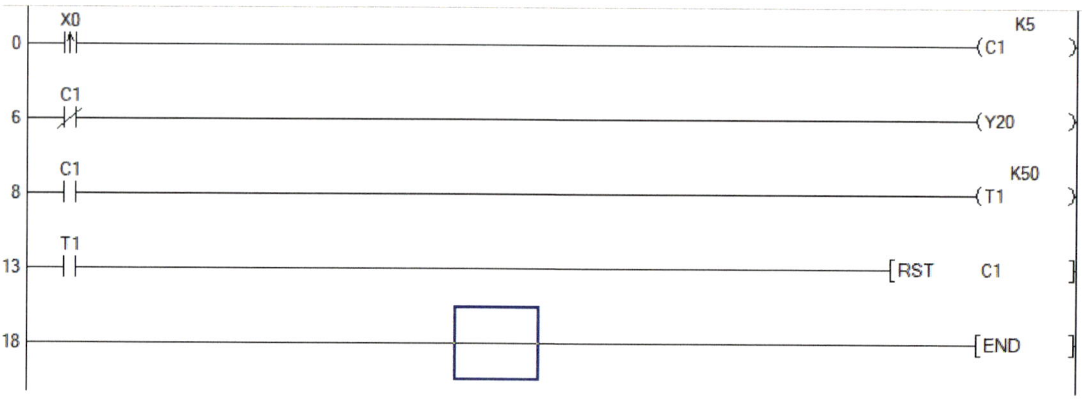